Help! My Computer Is Killing Me

MY COMPUTER IS KILLING ME

Preventing Aches and Pains in the Computer Workplace

DR. SHEIK N. IMRHAN

Taylor Publishing Company

Dallas, Texas

DEDICATION

I would like to dedicate this book to everyone who has had to endure the aches and pains caused by sitting in front of a a computer to earn a living. Without their efforts, many aspects of work in our society would come to a standstill.

Published by Taylor Publishing Company
1550 West Mockingbird Lane
Dallas, Texas 75235

Designed by David Timmons

Library of Congress Cataloging-in-Publication Data

Imrhan, Sheik N.
Help! My computer is killing me : preventing aches and pains in the computer workplace / Sheik N. Imrhan.
p. cm.
Includes index.
ISBN 0-87833-925-6
1. Human-computer interaction. 2. Human engineering. I. Title.
QA76.9.H85I573 1996
613.6'2—dc20 95-40154
CIP

Printed in the United States of America

10 9 8 7 6 5 4 3 2 1

CONTENTS

Acknowledgments

I would like to thank all those who read this manuscript or part of it during the development stages and offered suggestions for its improvement, especially Monica Steele for sharing her important experience from managing safety and ergonomics in the office and other sitting workplaces.

Thanks also to my daughters Sabine and Sayina and my wife Victorine for their sustained encouragement over the past six years.

INTRODUCTION

We are living in the information age. A large part of our daily business is conducted by the transfer of information. Traditionally, the office was responsible for this kind of transfer, and the typewriter was the most important piece of equipment. Information was typed on paper and stored in cabinet files and drawers. Then came the computer. The invention of the computer was a major breakthrough in the processing of information. It made the generation, storage, and retrieval of information easier and quicker. It revolutionized the nature of office work.

In their early days, computers were large boxlike, expensive structures. But they quickly evolved into small and powerful independent units that fit on office desks. This is why they became known as **desktop computers**. The term **personal computer**, or **PC**, is more popular now. The development of software (programs) further expedited their use in the office and other environments. Today, the great power, low cost, ease of use, and compatibility among different models of PCs have made them ubiquitous in offices and other computerized work environments. In fact, they have almost completely pushed the faithful typewriter into extinction. Today, typewriters in the office are considered inefficient antiques, and the modernity of an office is often determined by the quality and quantity of its PCs. But, ironically, it is this modernity that has brought tremendous suffering to those who use PCs—suffering in the form of aches and pains.

PC use has been expanding by leaps and bounds. In 1991 there were an estimated forty million PCs in the U.S. In 1995, the amount was estimated to be over sixty-five million. Projections indicate that by the turn of the century about 75% of all jobs will involve the use of computers. Many people now work at computer stations from their homes, and this trend is expected to continue as employers realize that many jobs do not require the physical presence of an employee at the traditional workplace. The use of computers at schools will also continue to proliferate as experts discover how to computerize teaching aids.

The desktop computer and its accessories have been accused of many evils since they replaced the typewriter in the 1970s. The computer screen (or monitor) has been accused of causing eye cataracts, malformed fetuses in pregnant operators, and miscarriages; but this fear has been largely quashed by lack of scientific evidence. In the 1980s, the National Institute for Occupational Safety and Health (NIOSH), the federal agency responsible for studying hazards in the workplace, conducted a detailed study of computer radiation. It concluded that the strength of radiation emitted by the monitor was very low and did not pose a problem to workers' health. Three other important studies in Canada, Sweden, and Finland also failed to find any clear link between computer use and adverse pregnancy outcomes. However, the matter has not been concluded. Controversy persists around other characteristics of radiation that need further investigation. A small group of scientists thinks that sudden strong surges of otherwise harmless types of radiation, called ELF (Extremely Low Frequency) and VLF (Very Low Frequency) radiations, can cause cataracts and reproductive problems. Traditional radiation measurements do not account for these surges.

Even if there is absolutely no harmful radiation from computer equipment, the lingering thought of its possibility may produce psychological stress great enough to contribute to a miscarriage. It is also possible that psychological stress from typical computer work (other than the fear of suspected radiation effects) coupled with prolonged sitting, which does not allow free flow of blood to the pelvic organs, may contribute to miscarriages. If you are in the later stages of pregnancy, your enlarged abdomen may restrict blood circulation if you sit in constrained postures. Therefore, it may not be considered unreasonable for you to ask your supervisor for a temporary transfer to a job that does not require prolonged sitting in one position.

Today the focus of health problems caused by computers is on aches and pains in the body. These problems are real, and they have always existed with desktop computer work. They were not given enough attention in the '70s and '80s, when radiation fear was rampant and overshadowed all other health concerns. But now they command as much attention in the working population and general public as almost any other work-related health problem. Scientific evidence indicates that these aches and pains originate from three main sources in the workplace: the nature of computer work, the design of the work environment, and the quality of lighting. To prevent aches and pains we should concentrate our efforts on these sources. How much effect they may have on you also depends on your physical fitness and vision. If we look carefully at who works at a desktop computer, we will see three kinds of work behaviors that cause aches and pains: prolonged sitting, intense reading, and repetitive finger motions.

Computer Work and Health Problems

At first many people thought that the computer would make work easier. Certainly, the coding and transfer of information were simplified and accelerated beyond imagination, and individual worker productivity and overall job production increased dramatically. But the toll on the worker's body was heavy. Listen to the complaints of employees who sit and work in front of computers all day. Many of them will tell you that their work is not as easy as it looks. Sitting takes a toll on their necks and backs, and the simple finger movements of using a keyboard can result in severe aches and pains. Their complaints should be taken seriously.

The office is now plagued by illnesses that were traditionally associated with highly repetitive or strenuous manual jobs found in manufacturing and mechanical maintenance and the service industries. Examples of these jobs or tasks are assembling parts, packing containers, buffing, grinding, sanding, polishing, hammering, meat cutting, using manual screwdrivers or wrenches, cashiering, and so on. The symptoms of these illnesses include muscular fatigue, muscular aches and pains, joint stiffness and pains, visual strain, inability to relax, mental fatigue, tiredness, and general tension. The occurrence of these maladies is so pervasive in the computer workplace that some people refer to them as a workplace epidemic. Many people feel as if they are fighting a slow, painful war against the computer. They are facing pressing health issues concerning computer work. Everyone who works in front of a computer should be interested in learning as much as possible about maintaining a healthy computer workplace.

Sources of Information on Computer Work Stress

There is a large amount of scientific information concerning the causes, effects, and prevention methods for computer-work-related illnesses. However, much of it does not seem to be reaching the affected computer users and the general public. The information is stowed away in technical journals and books that are difficult to understand. They are shrouded in highly technical language from the medical and engineering fields. There are a few nontechnical books that offer advice on how to prevent aches and pains caused by computer work, but they do not describe the problems in detail. They typically give lists of recommendations, often in the form of dos and don'ts, without explaining the underlying scientific concepts. Without a fairly sound knowledge of human anatomy and engineering, you would find it almost impossible to understand the scientific basis behind these recommendations, and some of the advice may not make sense to you even though it is sound. In this book, I explain the scientific aspects of computer-work-related problems in nontechnical language and provide advice on preventive methods.

I believe that even if you do not have a technical background, you can still be taught the science behind the aches and pains caused by computer work. With this kind of knowledge you would then be able to understand the reasons behind recommendations for preventing aches and pains while sitting and working and devise your own methods for relief and prevention where recommended ones prove to be unsatisfactory or where there are no recommended methods available. Few recommended solutions work for everyone or under all work conditions, and excellent workable solutions do not even exist for some problems. So you must be in a position to improvise. This book teaches you how to do that. It is a series of lessons on how to take care of your health in the computer workplace.

The lack of accurate information available to the general public and the huge gap in knowledge it creates are the main reasons these health problems are perceived as mysteries that are unsolvable and beyond the comprehension of the many computer users. This book circumvents that mysterious path. It has taken years of careful scientific research by different kinds of scientists and engineers to discover the facts about the aches and pains caused by sitting and working. But, as you will learn from this book, these facts can be learned fairly easily.

Because of the shroud of mystery, people tend to leave these problems to physicians trained to heal the body and ergonomists trained to design the workplace to prevent injuries and illnesses. However, computer users can help themselves prevent many of these problems and leave only the more difficult ones to the experts. The average computer user can recognize work conditions that contribute to or cause aches and pains and learn how to prevent many of them. But first, you must learn the principles behind these problems. This book will be your guide.

Knowing Your Problem Is the First Step to Solving It

After reading this book, you will be able to set aside much of your anxiety and frustration about what real computer-work-related illnesses are. Of course, you will not become an expert overnight in computer-workplace problems, but you will no longer have to rely on technical experts to teach you the basics of preventing aches and pains. To a large extent, you will be able to evaluate many important aspects of your workplace and even advise your employers about critical working needs (for example, the type of furniture, the arrangement of the workplace, the amount of desk space, and the quality of lighting). You will be able to participate directly in decisions that affect your health at work. Managers and supervisors who read this book will also be able to understand the problems of employees and communicate with them more effectively.

In this book, I have avoided the technical language that makes scientific reports incomprehensible to the general public. I have also tried to steer clear of the dramatizing or intimidating language of the news media. I think people would like to understand what their problems are without being terrorized about so-called computer epidemics, which, according to some popular news media reports, are likely to cripple people who work at computer jobs. This book takes a straightforward, factual look at the problems so that you can understand the implications of computer work without being scared of them. People who work at computers regularly, especially data entry clerks, word processors, and their supervisors, should definitely read this book. But it is a truly valuable tool for everyone who works at a computer. The material in it can answer most of the questions supervisors are faced with when trying to improve the workplace to prevent aches and pains. Other occasional users—engineers, managers, other professionals, and home users—will also find it to be a useful resource.

In addition to descriptions of aches and pains in your body, this book also offers solutions to the problems—solutions that you can try for yourself. These are the same kinds of solutions that highly qualified work-design specialists would recommend to workplace management for implementation at your workstation. I have explained the solutions in as simple a way as possible so that you can understand and implement them with little or no outside help. Simplification of highly technical material is not an easy task. To maintain accuracy, I have retained certain technical terms and concepts, but each has been defined the first time it is used and can also be found in the glossary at the back of the book.

This book begins with an account of the occurrence of aches and pains in your body from computer work, followed by a description of common types of computer work and a description of the structure and functions of the parts of your body that are affected. There are also explanations of how pains accumulate in your body—from sitting postures, finger motions on the keyboard, and intensive use of the eyes. Methods on how you can avoid these problems are described throughout the book.

Important points in each chapter are listed at the end under the section "Points to Remember." You may want to look over these points before reading the details of the text to get an idea of what is discussed in each chapter. You may also find it useful to read them after reading the whole book to reinforce your memory.

Benefits for the Home Office

Home computer users can also benefit tremendously from this book. You will learn some important methods of setting up a computer workplace and adapt them to the home-office environment. The way a computer workstation is set up in a home is similar to the way it is set up in an office. As you will learn later, many of the sources of aches and pains originate in the way you sit and work, whether it is in the traditional working office or in your home office.

Parents can gain the kind of knowledge from this book that will help them select appropriate work furniture and set up a comfortable computer workplace at home for their children. Parents will also gain the knowledge necessary to advise their children on the benefits of sitting properly—a lost art.

What You Will Learn in this Book

Here are just some of the things you will learn in this book:

- Why computer work is stressful, physically and mentally.
- Why constrained or unaccustomed postures can produce aches and pains in your body.
- How computer work and workstation design force your body into constrained postures.
- How certain sitting postures can prevent many of your aches and pains.
- How some furniture designs produce better sitting postures than others.
- How some methods of the arrangement of furniture and equipment produce better postures than others.

- How the speed at which you work produces aches and pains in your hands.
- How computer keyboard design strains your arms and hands.
- How poor workplace lighting affects posture.
- How poor lighting produces aches in the eyes, head, neck, and back.
- How your uncorrected vision produces aches in your eyes, head, neck, and back.
- How your glasses can contribute to aches in the neck.
- How exercise and work-rest regimens can help or fail to help you.

POINTS TO REMEMBER

- ❖ Desktop computer use will continue to rise rapidly—affecting about 75% of all jobs by the end of the century.
- ❖ The scientific community, except for a small number of researchers, agrees that the computer presents no danger from radiation.
- ❖ Computer work is a major cause of the rapid increase in reported cases of aches and pains in today's workplace.
- ❖ If you learn how your body develops aches and pains from computer work, you will know what to do to help prevent your suffering.
- ❖ This book is aimed at people who work at computers. It explains technical material in nontechnical language.

CHAPTER I

STRESS IN COMPUTER WORK

People who work at computers suffer from a variety of job-related health problems in different parts of their bodies. However, what seem like unrelated disorders can be better understood if they are categorized according to the systems in the body that they affect: muscular and joint aches and pains, vision problems, and psychological problems.

Evidence from Scientific Studies

You can determine how stressful a particular job is by considering the percentage of workers in it who suffer from work-related illnesses, and you can determine how harmful particular equipment is by comparing similar jobs with and without it. Both of these methods have been employed by experts in assessing computer work, and both indicate that computer work can be stressful and harmful to your health. Many scientific studies on computer-related health problems show similar results. It is not necessary to give all the details here, but the highlights of a few landmark studies emphasize the nature of the problems.

A 1980 study by NIOSH (the National Institute of Occupational Safety and Health) of newspaper offices and an insurance company in San Francisco revealed a high occurrence of musculoskeletal strain among employees who used computers. Eighty-one percent of clerical workers using computers reported that they suffered from neck and shoulder pains, while only 55% of those who did not use computers were affected by these pains. In addition, 78% of computer operators reported lower back pains, while only 56% of others reported such pains. In that study, a high occurrence of eyestrain was also reported by clerical workers and professionals who used computers. Of the clerical workers, 91% were afflicted, and of the professional workers, 78%. On the other hand, among clerical workers who did not use a computer, only 60% were affected.

Another study in Zurich, conducted by the Swiss Federal Institute of Technology, found that 50% of data entry clerks had muscle pains but only 23% of typists had such pains. The Department of Health in New Zealand found similar results in a large survey at a number of work sites. Muscular discomfort affected 37% of computer operators but only 26% of those who didn't work at computers; and eye fatigue affected 50% of those whose jobs involved computers but only 33% of those whose jobs didn't. Also, employees who took frequent, informal breaks from visual work were less affected than those who did not (42% versus 62%).

Confusion in Interpretation

The scientific community has now recognized the connection between computer work and health problems, but many questions remain unanswered. The causes and consequences of computer-work-related health problems are not easily understood, and they are not as clear cut as, for example, a fall from a ladder or a broken leg. A lot of confusion still exists about what problems actually occur in the body, why they occur, and what to do about them. In addition, the law seems to be unsure about how to interpret the relationship between computer workplace maladies and the nature of computer jobs. Employers are baffled about their manifestation in the workplace, and the medical profession often has difficulties in diagnosing the afflictions and identifying their causes. Many people are frightened because they must confront these strange sources of misery in their workplace every day.

People usually acknowledge that these problems are related to their work. They complain about the effects, such as fatigue and pains, but they are often not sure about what to do to prevent them. They seem to harbor the notion that this is what computer work is all about and that they have no option but to live with it. Supervisors and management are often of little direct help because they rarely possess the kind of knowledge and expertise related to these workplace problems. Few professionals in safety or management positions have had the rigorous educational background or training to solve these problems. But there is hope. Recent emphasis on workplace health and safety by OSHA (Occupational Safety and Health Administration, the federal agency responsible for enforcing safety and health laws in the workplace) has forced many employers to develop appropriate programs to deal with these new workplace maladies. Larger companies have, in turn, sought educational and training opportunities for their safety and health professionals so that they can deal with these problems. Many qualified professionals have also created **ergonomics** consulting companies dedicated to solving workplace health problems. (Ergonomics is the science of coordinating physical working conditions with the capacities and requirements of the worker.) Unfortunately, it seems as if many employers are motivated to maintain safe and healthful work conditions only when they are under threat of heavy fines for violations of OSHA's laws.

Stress and Strain

The terms **stress** and **strain** are commonly used when describing work-related illnesses, but there is some confusion about the way they are used. In correct usage, stress refers to some undesirable condition that affects you adversely, and strain refers to your body's reaction to the stress. This definition helps to separate the cause of the problem from the problem itself. If, for example, your workplace is noisy and the noise causes you to have a headache, then the noise constitutes stress and your headache is the resulting strain. Some examples of stress and its corresponding strain caused by computer work are given below to further illustrate the difference between these two terms.

Stress	Strain
Awkward sitting posture	Excessive pull tension in muscles and mechanical pressure in the joints
Fast rate of keying	Pull tension in the finger muscles, and friction on the tendons inside the wrist
Insufficient rest	General fatigue or headache
Dim lighting	Fatigued eye muscles or fatigued optic nerve

Illness and Injury

Two other terms that are often misused when describing computer-related health problems are **illness** and **injury**. Injury refers to instantaneous damage by a sudden force, for example, a cut or a broken arm. Illness, on the other hand, implies an adverse change that develops gradually over a period of time in the body. The term **occupational** refers to the fact that the illness or injury is related to or caused by one's job. The aches and pains that people who work at computers suffer from are mainly occupational illnesses, not injuries. However, the terms injury and illness have been used so interchangeably that the distinction has become blurred. One particular occupational illness from computer work has become widely known as an injury—repetitive strain injury, also known as repetitive motion injury. Because of the widespread use and acceptance of this term, I will use it in this book.

Otherwise, I will refer to cumulative strain problems in the body as illnesses or disorders.

Musculoskeletal Problems

Musculoskeletal aches and pains refer to the effects of strains in the muscles and joints and their associated structures due to physical stress. In computer jobs that require sitting for long periods of time, the problems arise mainly from two sources. The first is the steady, sustained contraction of certain muscles when the body stays in a fixed seated posture for a long period of time. Aches and pains are felt mainly in the back of the neck, the shoulders, and the lower back. In some situations, fatigue is also felt in the legs. The great speed at which the fingers work on the keyboard is the second source. The aches and pains are due to the gradual wear and tear on the tendons of the finger muscles and pressure on a crucial nerve, the median nerve, that passes through the wrist.

Vision Problems

Visual fatigue is also known as eyestrain. In the computer workplace it is caused by prolonged and intense use of the eyes when reading from a display, such as the computer screen, a document, or the keyboard. In most cases, the screen is the culprit. The symptoms of eyestrain are numerous and depend on the severity of the strain. Blurred vision, double vision, or difficulty in focusing is common. Watering or soreness of the eyes has also been reported. Headaches usually occur as a result of struggling to cope with visual difficulties in reading. In many cases, headaches may be accompanied by irritability, general tiredness, or dizziness.

Eyestrain is caused primarily by overuse of certain muscles in the eye. These muscles are involved in moving the eyeball, controlling focus, and controlling the amount of light entering the eyes. The mechanisms are explained in chapters 14 and 15.

The decreased capacity of the eyes in people over forty-five years of age, the use of bifocals or trifocals, poor lighting, and glare are the major factors that foster visual problems in the computer workplace. These problems can also lead indirectly to poor body posture and, as a result, musculoskeletal strains. People usually don't associate vision problems with aches and pains in the muscles and joints of the body, but the relationship does exist. For example, when the computer screen is too far away, you may have to lean forward to read from it; or, if overhead lights are reflected off the screen, you may be forced to bend your neck sharply to see around the glare. When either of these situations is prolonged, your muscles and joints will be subjected to considerable strain, eventually leading to aches and pains.

Psychological Problems

Psychological problems that are experienced in the computer workplace are similar to those in other types of modern work environments. The stress arises from a variety of work conditions that are not completely understood but put pressure on you. Here are some of those stressful conditions:

- Pressure to work quickly
- Large volume of work
- Regular and tight deadlines
- Repetitive and monotonous tasks (as in data entry)
- Inadequate breaks
- Limited opportunities for interaction with others
- Physical barriers (work cubicles)
- The feeling of being enslaved and controlled by the computer
- Computer monitoring of performance by management

Psychological reactions or strains affect our feelings and our behavior. Common symptoms are anxiety, anger, general tension, irritability, confusion, depression, and boredom. Psychological strain has also been referred to as mental strain. Some of the body's reactions to psychological stress are manifested physiologically, for example, by general fatigue, sleep disturbance, ulcers, headaches, or irritability. These are known as **psychosomatic reactions**, and they are common in computer jobs where the stress is prolonged over months or years.

The use of the computer has also introduced a new source of psychological stress—the continuous monitoring and pacing of worker productivity by management. According to a 1993 issue of *MacWorld* computer magazine, as many as twenty million people at that time were subjected to computer monitoring in U.S. workplaces. In some types of computer jobs, the speed and volume of work performed, as well as accuracy, can easily be monitored by the computer system. Management may then use the results of these measurements to determine pay and other rewards. From an employee's perspective, this is stressful. To know that every move you make at work is watched through the eyes of a computer by the people who determine how much you should earn or whether you should have a job can be very stressful. Along with the lack of privacy, it is definitely not a welcome work condition. It bears strong resemblance to the watchful eyes of Big Brother in the fictitious society created by George Orwell in his masterful novel *1984*.

Vagueness of Names for Computer-Work-Related Illnesses

The scientific and medical professions often have difficulty in clearly distinguishing the symptoms of computer-work-related disorders or pinpointing their causes. For this reason, there are many terms that refer to the same illness and even the most popular terms used to describe these disorders are vague and confusing. Here are some of these terms:

- Psychological strain
- Mental strain
- Cumulative trauma disorders (CTDs)
- Repetitive motion injuries (RMIs)
- Repetitive strain injuries (RSIs)

Cumulative trauma disorder (CTD) is one of the most popular terms used for describing musculoskeletal problems arising from physical overexertion. It refers mainly to damage to the muscles, tendons, ligaments, joints, and nerves in the body from gradual accumulation of the effects of mechanical forces on them over a period of time. The mechanical forces include pull tension in muscles, tendons, or ligaments and compression, sliding, or twisting in joints. Sliding and twisting are also known as shearing and torsion forces, respectively. In this book, these forces will be referred to collectively as **mechanical strain**.

The time it takes CTDs to affect your body may be weeks or years. The exact time depends on a number of factors, the most significant of which are the severity of the strain, the amount and distribution of rest time within your work, and your health and fitness levels. The strain may have little effect at first but can become severe if your body does not get enough rest.

Repetitive motion injuries (RMIs) are specific types of CTDs. They refer to a variety of ailments in the musculoskeletal system, concentrated mainly in the neck, shoulders, and hands and caused by the gradual accumulation of strain from repeated hand motions throughout the workday. The emphasis here is on repeated motions. The accumulated strains may affect the muscles, tendons, and nerves that are involved in muscular contraction. In computer work, RMIs are usually severe in the area from the wrist to the fingers, since the fingers are the only parts of the body that move repetitively. You may suffer from mild aches to intense pains in your hands and wrists, loss of control of fingers, decreased hand strength, loss of grip, or numbness. **Repetitive strain injury (RSI)** is just another term for RMI.

The History of Cumulative Trauma Disorders

Cumulative traumas from work were recorded as early as the 1700s by the Italian physician Bernardini Ramazzini, and their diagnoses and medical treatments were fairly well established by the middle of this century. However, they were not studied systematically in the workplace until fairly recently. In 1713 Ramazzini pointed out that constant unnatural postures, unnatural body motions, and repetitive hand motions were common in the workplace, and that they were responsible for debilitating illnesses among workers. Today, almost three hundred years later, stresses in the workplace are not much different. The types of work have changed, machines have evolved into more efficient tools, and the workday is much shorter. But the same types of bodily stresses and strains persist. In many work environments, especially the office, work is still characterized by fixed postures (such as sitting), repetitive motions of some body parts, and psychological stress.

Even though Ramazzini was not referring to sitting and typing in the office (the typewriter keyboard was not invented until 1873), his observations were similar to symptoms of cumulative trauma disorders rampant in the computer workplace today. Your body is subjected to muscular strains from work that may seem easy and harmless, but these strains accumulate to harmful levels over a period of time. Work that is repetitive or is maintained for long periods of time without rest produces great strains in muscles and joints. Pressing a key on an electronic keyboard, for example, requires a relatively weak force—only a few ounces. But when those weak forces are applied anywhere from 17,000 to 20,000 times per hour, as is common in some types of computer jobs, something in the body is bound to give. The cumulative strains from these motions wreak havoc on the muscles and tendons that move the fingers. Repetitive work without enough rest gradually breaks down the body's recuperative mechanisms, even if the muscular efforts are not strong. This is the basic problem with office computer work.

The body posture that you adopt at your computer is almost identical to that of the traditional typist's. Few people would consider it stressful. But it is maintained for long periods of time, and the muscles (mainly those in the neck, shoulders, back, and legs) required to maintain this posture simply do not get enough rest or opportunities to contract and relax rhythmically. Small, imperceptible strains, therefore, gradually accumulate until they become unbearable.

You need to understand how computer work can be made easier for your body so that your health does not become impaired by it. Of course, your goal is to accomplish this without resorting to too much rest at work, because you still want to maintain productivity. This is no easy task, but proper design of the computer workplace and equipment has been the most effective method for achieving satisfactory results. This has been done through the application of ergonomics.

POINTS TO REMEMBER

- ❖ People suffer from strains that accumulate over a long period of time—cumulative trauma disorders (CTDs).
- ❖ CTDs in the computer workplace are not well understood. This is reflected in the great variety of names used for the same illnesses.
- ❖ The causes of CTDs from work were reported almost three hundred years ago by the renowned Italian physician Ramazzini, but they were not systematically studied until a few decades ago.
- ❖ The main disorders (illnesses) are musculoskeletal, visual, or psychological.
- ❖ Musculoskeletal problems are due to either sustained contraction of muscles from maintaining a sitting posture or repetitive contractions from moving your fingers rapidly.
- ❖ Visual problems or eyestrains are due to prolonged and intensive use of the eyes for reading. Poor quality or level of lighting exacerbates these problems.
- ❖ Psychological problems are due mainly to the constant pressure for maintaining a fast pace of monotonous work and a lack of control over your work.

CHAPTER 2

COMPUTER WORK IS HARD WORK

When the computer was introduced in the office and other workplaces, few suspected that it would be associated with the kind of aches and pains it is blamed for today. Unlike a lot of heavy industrial equipment, the computer had no large, heavy, or fast-moving exposed parts, it spewed no dangerous chemicals into the environment, and it emitted no loud noise that damaged the delicate structures in the ear. Moreover, computer work was not associated with heavy muscular exertions of the body, such as lifting, lowering, carrying, pushing, and pulling heavy loads or powerful manual forces, like pressing down, gripping, turning, and pinching. Except for fears about the harmful effects of radiation emission from monitors, computer equipment was perceived as being relatively harmless. So it's a small wonder that some people still cannot believe that it is responsible for their constant aches and pains at the end of the workday.

Contrary to general opinion, computer work is hard work. Many parts of the body work strenuously. The eyes are involved in viewing the screen or documents, and the fingers move like pistons over the keys, faster than the eye can follow. At the same time, the muscles in the back, shoulders, and legs must maintain a constant state of contraction to keep the body in a stable position. They act like guy ropes attached to a ship's mast to prevent it from falling. Jobs that require this kind of sustained contraction from sitting and working may be as strenuous as jobs requiring heavy muscular work. Intense concentration for maintaining accuracy and the psychological pressure associated with meeting deadlines add to the burden.

Hard Work and Cumulative Strains

In general, the effects of mechanical or psychological strain on your body can disappear with adequate rest. But without rest, the effects will accumulate day after day. This accumulation may not be noticeable at first, but as the body's recuperative mechanisms become overwhelmed, you begin to feel the resulting aches and pains. From that point on, you will always be conscious of the strain. For example, working while bending your neck sharply induces strain in those muscles. Tightness and mild fatigue may be felt after a short while but will disappear if your neck gets enough rest. Otherwise, the tightness and fatigue can translate into persistent aches and pains that will be felt even when you are not working at your computer.

It may take months or years for the symptoms of debilitating cumulative strains to appear, and, because of the gradual nature of these disorders, their causes and time of onset are difficult to pinpoint. This is one of the reasons why, until recently, managers, physicians, and many other professionals doubted that these aches and pains were associated with overexertion from computer work. They could not pinpoint a specific incident responsible for these disorders or a specific time they occurred. Management and other administrators were accustomed to thinking in terms of injuries caused by a single event, such as a crushed finger, a broken leg, or a chemical burn. They were not accustomed to thinking in terms of cumulative disorders, in which no single harmful incident could be identified as the cause. In fact, traditional safety departments were concerned mainly with hazards that could be detected with the senses or with instruments and could be directly associated with injuries or illnesses. What was not understood was often ignored or denied—that was the fate of CTDs.

People have always known that prolonged sitting and working in front of the computer were associated with aches and pains, but they did not always understand exactly what was happening in their bodies or why it was happening. In fact, many thought that their bodies were to blame. They thought that their bodies were not strong and resilient enough to perform hard work. Many also believed that, after a good night's rest, the aches and pains would disappear. But sadly, many of them discovered that a good night's rest was just a temporary relief. In fact, in one particular type of repetitive motion injury, **carpal tunnel syndrome**, pains in the wrist and hand usually affect the sufferer by night as well as by day, and sleeping often becomes a struggle between a tired body and an activated brain sending out pain signals to the hands.

Do Workers Fake CTDs?

In the U.S. the Bureau of Labor Statistics (BLS) conducts an annual survey of occupational injuries and illnesses. Among this data for illnesses are those for cumulative trauma disorders (CTDs). It shows a remarkable upward trend in the incidences of CTDs—from 18% of all reported occupational illnesses in 1981 to over 66% in 1993, with 302,600 workers reporting the disorders in 1993. The jobs with the most cases are motor vehicle and equipment manufacturing, aircraft and parts manufacturing, and clothes manufacturing.

Some people claim that the sharp rise in reported cases of CTDs in recent years is due largely to dishonest workers faking these illnesses (malingering). No doctor or instrument can measure pain, so it would not be impossible for people to deceive doctors about the existence or severity of pain. Undoubtedly, some workers have falsely reported CTDs, but I doubt the practice is as widespread as claimed. There is no conclusive, objective evidence to indicate that malingering is widespread. What are the reasons, then, for the sharp rise in reported cases of CTDs? There are three credible answers:

- Workers' compensation boards now classify CTDs as compensable illnesses. This means that employers or their insurance carriers must now bear the costs of medical treatment for these illnesses if they are work related. Before the early 1980s, work-related CTDs were not compensable, so it is likely that many workers just did not bother to report CTDs to their employers.
- There was a perception among workers that they would be considered worn out and unproductive if their employers knew of their illnesses. Accompanying this perception was the fear of being laid off (ostensibly for other reasons). Many affected workers, therefore, kept their illnesses a secret. This fear still exists today, but it seems to be less common because employers cannot ignore reports of suspected CTDs, making workers more likely to report their work-related illnesses.
- Because of the impact of education about work-related illnesses, workers are now better able to recognize when their symptoms are work related and report them as such.

Don't Blame the Computer—Blame Computer Work

The computer is the end product of a long line of machines that were designed to aid mental effort. It arose out of the need to generate, store, transmit, and process large volumes of data quickly. So far, it has performed beyond our wildest dreams and is expected to soon be able to do things beyond our imagination. Perhaps it is blamed for so many illnesses precisely because it is such an amazing machine. We seem to have an almost natural apprehension for strange or new machines and a tendency to reject those that change the way we are accustomed to doing things. This attitude tends to be intensified when we believe that those machines can replace us at our jobs.

Machines have been the key to the progress of civilizations. They were invented to aid the muscles and the brain and to enhance the power and efficiency of physical and mental work. They have enabled the production of goods in large quantities in short periods of time and, in doing so, have improved the quality of life to levels unimaginable a century ago. But beneath the canopy of their glory, these achievements have left a trail of suffering and sadness.

The proliferation of machines and industries, especially during the period of the Industrial Revolution, was accompanied by a spate of injuries to the body. Many machines and jobs demanded so much from people that parts of the body were often strained, burned, crushed, mangled, poisoned, severed, or otherwise destroyed while using them. The offending machines were often fearsome to the eyes and ears. Few people welcomed the sight of rotating blades, an exhaust spewing flame and smoke, or the sound of deafening noises. Many such overtly dangerous machines still exist today, but the computer is not one of them. Yet the level of suffering that the computer has been blamed for matches the levels for many of these dangerous machines.

The answer to this paradox is that there is a clear difference between a machine's characteristics and the way in which it is used. The computer itself is not hazardous, but the way it is used may create conditions that are hazardous to your health. The nature of the work for which the computer is used is often the real culprit. Data entry, for example, requires such large quantities of information to be entered into the computer via the keyboard every day that your fingers must work like the pistons of a racing car. Fingers strike the keys at speeds as great as 17,000 key strokes per hour—almost 5 per second! Consider this amount over a year:

- 102,000 in a day (assuming 6 hours of key stroking per day)
- 510,000 in a week (5 days)
- 25,500,000 (twenty-five and a half million) in a working year of 50 weeks!

The use of the mouse as an input device has eliminated some keying motions and is expected to eliminate considerably more as technology continues to develop, but it is not safety-proof either.

The arrangement of furniture and other work equipment also contributes to these stresses and strains to a greater extent than many people realize. Poor positioning of the keyboard, the computer screen, or a document, combined with poor furniture design, can easily force you to sit in constrained, unnatural, and stressful postures—postures that produce pains in the body. In many types of computer work, a few small, localized parts of the body work throughout the day and overload themselves and you often have little opportunity to rest. Data entry and word processing fall into this group. They involve tremendous movements by the fingers on the keyboard while the rest of the body remains motionless on a chair. Drafting, design engineering, and financial analysis are similar well-known stressful jobs.

Such work developed from the impetus of society to produce large quantities of goods in short periods of time. Society's methods were refined by the invention of the scientific method of work, popularly known as **Taylorism**.

POINTS TO REMEMBER

- ❖ The gradual nature of cumulative strains makes it difficult to identify a time of onset.
- ❖ CTDs are now classified by workers' compensation boards as a compensable illness.
- ❖ The computer itself is not harmful to your health; the way it is used and the nature of computer jobs are the real causes of health problems.
- ❖ The muscles that contract and the joints that bend are under considerable strain when you maintain a sitting, working posture.
- ❖ A typical data entry operator makes 17,000 keystrokes per second, with the fingers, or 25,000,000 per working year.

CHAPTER 3

WORK METHODS AND COMPUTER STRESS

Computer work is one kind of mass production. Mass production involves the production of goods in large quantities by specially designed methods in relatively short time spans. Mass production has radically changed the way people work and increased work rates and productivity markedly, but it has also had some harmful results.

At one time, one craftsman produced an item by himself. The process typically involved many steps, and the craftsman was skilled in all. With the advent of mass production, the single multiskilled craftsman was replaced by many workers, each of whom performed only one or a few steps. Each of these workers was a specialist, so to speak. Such mass production methods increased the rate of production dramatically. Ten specialists could produce more items than dozens of craftsmen who possessed overall skills. But mass production had its problems.

Mass production brought people together in large numbers, under the same roof, at the same times of the day, with the same objective: to manufacture items according to predetermined specifications. Instead of working for themselves, as was common in the pre-industrial days, people worked for others—for the owners, who were far removed in body and mind from the drudgery of the mass-production workplace. Workers were compelled to work at set times during the day for set durations at specific jobs. A single worker, or a team, had to work according to a definite production plan. There was little room for the variation enjoyed by the craftsmen of years gone by. The work of any particular person was integrated into the work of others, and that individual lost the satisfaction of seeing the fruits of his own efforts.

With mass production, work also had to be synchronized for efficiency, especially when it involved tasks that were sequential in nature. One step had to be completed before another could be started. Timing of the various steps was, therefore, crucial to the smooth running of the system. Workers were expected not only to perform their tasks with minimal errors but also to maintain a certain speed. Otherwise bottlenecks arose, and work was then either slowed down or stopped. This obligation to finish a task by the tick of the clock added more weight to physiological and psychological strain among workers.

Loss of Freedom and Control

It was under this type of work condition that many individual freedoms in the workplace were lost—the freedom to choose your own job, your working hours, your own pace, the amount of work, and your rest breaks. Computer work is a modern example of work that binds you to these strict work rules. In certain cases, you may be able to choose a certain job freely, but you can hardly expect to choose the exact method by which it must be performed. For example, you may choose to work as a data entry operator, but you will not be able to choose the speed at which you must work, the amount of work to be completed in a given day, or the sequence in which you should perform several assignments. You will also have little choice in the number and duration of rest breaks or the exact time of the day when they should be taken. Even the furniture you use at your workstation (considered to be the source of most aches and pains) will, most likely, not be of your choice. These limitations in the choice of your work conditions are part of what defines how stressful, physically or mentally, your job can be.

Social Isolation

Traditional, noncomputerized office work was not as dreadful as the work described above. It provided opportunities for relieving boredom and its ills. And since it was not as intensive as computer work, there was more interaction among office workers. The need to feel part of a larger social group, even for a short period during the workday, is important to us all. However, in many computerized offices each person is enclosed in a small space or cubicle that constitutes a separate workstation. This means many people are physically isolated and deprived of natural, human social needs. Those who favor this kind of workplace design point out that the isolation prevents distractions from the activities of other workers and that people perform adequately under these isolated conditions. However, although job output may be adequate, job satisfaction may diminish, making workers likely to suffer from psychological strain.

The Scientific Method of Work

As early as the eighteenth century, Adam Smith, considered to be the founder of modern economic science, wrote that productivity could be greatly enhanced by subdividing work into separate steps. But it took more than one hundred years for this idea to be put into practice systematically by Frederick Winslow Taylor and the husband-and-wife team of Frank and Lillian Gilbreth. The basic idea was to break down each task into a number of small, discrete steps and to exhort people to perform those steps as quickly and efficiently as possible. These small, efficient steps would then add up to an efficiently completed task. This philosophy and approach to work later became known as **Taylorism.** Gradually machines were developed to enhance this kind of specialization. Machines perform better than people at tasks that are repetitive and predictable. Taylorism was applied in its most intense form in the industrial assembly line that was pioneered by Henry Ford.

Today, Taylorism has been almost universally applied to industrial production methods. As a result, items are produced much faster and in larger quantities than before. The demand for this availability of goods in adequate quantities has been the driving force behind mass production and the ensuing improved living standards.

Taylorism emphasizes motion economy. Body motions that do not contribute to the performance of a task are deemed wasteful and are eliminated by changing the way the task is performed. This change is part of what industrial engineers call job redesign. Though Taylorism enhances work efficiency, it also produces serious side effects. These effects were not given much attention until fairly recently when people began to understand how fatigue occurs in the human body at work. In many cases, motion economy methods are aimed at reducing physical work by using only that part of the body that is necessary to perform the task. The other body parts remain practically immobile. However, they are not at rest as many people thought. In fact, they are working hard. They become stabilizers for the body, keeping the body in its working posture. They are under sustained static muscular contraction. In other words, both the actively moving and the immobile, stabilizing parts of body are stressed during work.

Workers have become too specialized, so to speak. They are required to perform the same operation repeatedly throughout the workday. Such intense repetition and monotony eventually led to boredom because the creative centers of the brain are not stimulated enough. Also, in many jobs, work is concentrated in one or a few parts of the body, which become quickly overloaded. Many computer jobs fall into this category. Even before the introduction of computers in the office, many jobs were already understimulating and boring. The introduction of the desktop computer has simply aggravated the situation. It has intensified the application of Taylorism and reduced the variety of work people perform. It calls for more intense mental and visual work, less social interaction with coworkers, less use of some parts of the body, and greater speed of work.

Unfamiliarity with New Equipment

Although the benefits of technology tend to remain in our memories for a long time, the suffering and sadness it brings often dies with the sufferer. In the modern workplace there is little time for studying new tools or new methods of work. The pressure for companies to produce the best quality items at the fastest pace has generated a wave of technological improvement, new work methods, and new methods of management within a short span of time. New products appear in the marketplace, then disappear in the blink of an eye. They are either modified for improvement or replaced by other new products. This is true of the computer industry, and the rapidity of change is often another source of safety and health problems among workers.

After studying the effects of a wide variety of equipment in the workplace, work design experts concluded that we make fewer errors and work more efficiently with tools that we are accustomed to, as long as they are properly designed. Many tools that have been in the workplace for a long time have undergone gradual improvement or become used more effectively as workers become more familiar with them. The computer is a relatively new tool, and perhaps its imperfect design and our lack of familiarity with its work are at the root of many of its problems. New equipment often has unknown hazards that are seldom identified until the equipment has been in use for some time. This is a critical time, like the warranty period of a new car when design flaws are discovered or the first year after marriage when couples realize their incompatibilities. It is the time when we must monitor the computer-human interaction carefully and try to reduce its adverse health effects on workers. Although the rapidity of change in the computer industry and in computer work sometimes seems too great to keep up with, we must not be daunted by the magnitude of the task.

The hammer is one of the best examples of a tool that has evolved slowly. Its handle is comfortable to most people and its design looks simple. But this simple design evolved from the crude stone ax used by cave dwellers thousands of years ago. Its oval cross-section, texture of surface, length, weight, mass distribution, and location of its center of gravity are all features that gradually developed to fit the shape, size, and strength of the human hand and conform to human work needs. Fortunately for us, the evolution of machine design and work methods has been accelerated by the vast increase in knowledge and improvements in technology. So we don't have to wait thousands of years for any type of work equipment, such as the computer, to evolve to suit our work needs. But we must always remember that this acceleration can easily produce harmful interactions.

The desktop computer and its peculiar workstation have been around for only about two and a half decades, and their health problems are only now beginning to be understood. We must, therefore, continue to make a concerted effort to study and understand the computer workplace to get the most out of computer work and prevent the occurrence of occupational illnesses associated with it. We should not panic or allow doomsday prophets to scare us. What is true for equipment is also true for work methods. Computer work methods are not yet well understood. The harm caused by rapid finger work and constrained sitting postures is still not yet fully appreciated, and the design and arrangement of computer-workplace furniture has yet to be perfected. Computer work has simply not been around long enough for us to understand fully how it strains the body.

A Look in the Wrong Direction?

Until the ergonomic approach was applied to the design of the computer workplace and the nature of computer work, we seem to have been looking in the wrong direction for explanations to the health problems people were experiencing. There was, and still is, a tendency to look for weaknesses—physical or psychological—in workers and blame them for these health problems. This negative attitude has nourished, and has been nourished by, the proliferation of health clubs and so-called work hardening clinics. These clinics offer a kind of rehabilitation therapy for workers recovering from work-related injuries or illnesses by having them perform simulations of their work in order to eliminate or minimize strains and injuries on the job.

Although both health clubs and work hardening clinics have useful roles to play in maintaining health and reducing illnesses and injuries among people, they may inadvertently function as outlet valves for employers who do not want to look inside their workplaces for conditions that cause or contribute to illnesses and injuries. These two institutions may also inadvertently give you a false sense of security by making you feel as if participation in their programs guarantees your health and safety at work. They may also give the false impression that your health problems lie in your body and not your workplace.

Today, we have sufficient scientific information to conclude that when workers suffer from injuries and illnesses work conditions should be one of the first things examined. In the computer workplace, these conditions include the design of computer furniture, the arrangement of furniture and computer equipment, and the nature of the work. Ergonomic studies have shown that none of these aspects of computer work can be considered in isolation; the computer workplace must be treated as a whole system. A change in one aspect is likely to affect the others. If such considerations are not taken, then solving one problem may create new ones.

POINTS TO REMEMBER

- Mass production was responsible for workers specializing in jobs.
- Although mass production improved our material standard of living, it robbed us of certain freedoms, such as individual choice in the way we work.
- Taylorism, the scientific method of work, developed out of mass production.
- Even the parts of the body not in motion sustain strain and are susceptible to work-related illnesses.
- Rapid changes in equipment and work methods do not allow us enough time to examine features of work that are harmful to our health.
- Any time a work-related illness occurs, the design of the workplace, the equipment, the environment, and the nature of the work must all be examined.

CHAPTER 4

ERGONOMICS—A NEW SCIENCE TO THE RESCUE

Just after World War II, a relatively new science called ergonomics evolved out of the need to design jobs, equipment, the workplace, and the environment to prevent work-related illnesses and injuries and enhance work efficiency. Ergonomics (also called human factors or human factors engineering) evolved as a distinct discipline in the United States and Europe. It was originally aimed at solving problems in the military, problems that were made obvious during the war when military personnel found themselves lacking in certain physical or mental capacities to operate or maintain important equipment. Ergonomic methods that were developed for the military were later applied successfully to industry, especially in manufacturing. The most recent area of application is the computerized office.

Before the middle of the twentieth century, occupational illnesses were virtually unrecognized as illnesses of the workplace. It was common to ascribe the causes of injuries and illnesses to workers. The workplace and its equipment were often treated as harmless entities, and accidents were often ascribed to human error. This way of thinking took the blame away from poor equipment design or work methods. However, the science of ergonomics has shown that these attitudes were biased. Many so-called human errors are really the results of people's inabilities to perform a task that was poorly designed or their difficulties in using poorly designed equipment. Studies in ergonomics have repeatedly proven that many incidents labeled as accidents could have been prevented by smarter engineering designs. Another important achievement of ergonomics, and one that is more relevant to computer work, was the recognition that many cases of aches and pains were the result of cumulative traumas caused by poor workplace design, furniture design, or work methods. Computer work is the most recent of this stressful type of work.

The Philosophy behind Ergonomics

The basic philosophy behind ergonomics can be summarized in a simplified form as follows:

> *Work requires inputs from human functions, and each human function, such as vision, hearing, balance, strength, endurance, and so on has a limit. If a job must be completed successfully, without harm to human health, then its demands must not exceed the limits of human functions; otherwise the job, the equipment, or the work environment should be redesigned to conform to human capabilities.*

Ergonomists refer to this approach to designing things for people as fitting the task to the person. In computer data entry, for example, you should not be expected to enter data via a keyboard faster than your fingers can press the keys, sit on a low seat and raise your arms uncomfortably, or read from a document under dim lighting. The expected data-entry rate should be set within the speed limits of your finger motions; the seat height should correspond to the length of your lower leg; and the workstation should be illuminated to suit your vision. Until a few decades ago, engineers often designed equipment without carefully considering the functional limitations of the people who use that equipment.

The practice in design was to make equipment technically correct, and if you were unable to use it properly, then you would be considered deficient in some way. You would then be given training to use the equipment or considered unqualified for the task. This approach flourished when engineers were uneducated about the roles of humans in a work system and, unfortunately, was widespread until only a few decades ago.

There have been too many situations in the workplace where providing training or resorting to some sort of personal aid to improve human capacity has been the first step in trying to solve problems rooted in poor work design. This kind of approach can impose too much pressure on you because you will be expected to maintain a heightened level of performance throughout the workday.

The Impact of Ergonomics in the Workplace

Ergonomics has now spread to all aspects of the workplace. In fact, wherever people work, either alone or in groups, ergonomics has a role to play. The role becomes more important when the conditions are likely to produce a lack of fit or mismatch between a person's capacities and the demands of the work. Wherever the human body experiences difficulties in performing work, ergonomic solutions are appropriate. It is natural, therefore, for ergonomists to take the lead in seeking solutions to the problems of aches and pains caused by office work. Throughout the rest of this book, you will read about the ergonomic solutions to musculoskeletal aches and pains from work in the computerized office.

Another important reason for using the ergonomic design approach concerns your moral and legal rights to gain employment. One of the best ways to make people employable is to design work so that almost everyone, if not everyone, can perform it satisfactorily. Otherwise, there will be a tendency to classify some people as unqualified and eliminate them from job considerations. As all employers should know, this can lead to serious legal problems. A recent U.S. law, the Americans with Disabilities Act (ADA), expounds the same principles. The ADA seeks to make disabled people employable by mandating employers to redesign their workplaces, equipment, or methods of work, if necessary, and as long as it does not cause the employers undue economic strain to render disabled people employable. For example, an employer can no longer deny a wheelchair-bound person a job because the passages in the workplace are too narrow for the wheelchair. Instead, the employer must widen the passages to accommodate the wheelchair. Ergonomic principles are now used to expand the job market for people with disabilities. Unfortunately, the practice of considering some people, not necessarily disabled ones, unfit for a job (that was poorly designed) still flourishes in some areas.

Ergonomists and other scientists have studied the relationship between computer work and computer-related illnesses extensively over the past two decades. They have made considerable progress but are still unable to answer many questions. However, these questions serve to emphasize the difficulties and challenges in trying to solve the relatively new problems created by the computerization of the office and other workplaces. The fight against computer-work illnesses has just begun, and it must involve not only ergonomists and physicians but also management and workers.

POINTS TO REMEMBER

- Ergonomics deals with designing things so that people can use them effectively and safely.
- To design properly for people, we must match the demands of the task with people's capacities.
- Your body is subjected to strains when job demands exceed human capacities.
- You can prevent a lot of strain by redesigning your work and work environment to match your functional capacities.

CHAPTER 5

TYPES OF COMPUTER WORK

The computer, as well as the desk, chair, document holder, and other accessories necessary to perform computer work are usually called the **computer workstation**. The term workplace is often used in place of workstation. The computer workstation is the immediate work environment, and, in many offices, it is partially enclosed by walls that form a cubicle.

Not All Computer Jobs Are Equally Stressful

You may have heard some people saying that working with a computer is harmful to your health while others say exactly the opposite. One of the reasons for this difference in opinion is that people are often talking about different types of computer jobs. To clear the cloud of confusion, we should learn what the different types of jobs are and how they can stress human functions.

Knowing how your body interacts with the computer can help you to assess how strenuous your job is or how much stress a particular task imparts. This knowledge may even help you determine whether you would like a particular computer job as a career. It may also help management organize the workplace and interpret stress-related health complaints from workers more accurately.

Stress from your relationship or interaction with a computer occurs in three different ways:

- Sitting in a chair in a particular posture.
- Reading from the screen, the keyboard, or a document.
- Using your hand on the keyboard, mouse, or another input device.

Each of these types of interaction imposes stress on the body in a different way, but not all jobs involve the same level of stress. Sitting stresses mainly the musculoskeletal structures in the neck and trunk, reading stresses the eyes, and keying stresses the hands and wrists. The characteristics of a computer job determine which part or parts of your body will be stressed as well as the level of stress.

Different Jobs and Different Stresses

There are numerous types of computer jobs. You need to learn their characteristics to understand how they affect your body. For example, you need to know how similar or different the various jobs are and what parts of the body are affected by them. The late Etienne Grandjean, former director of the department of ergonomics and hygiene at the Swiss Federal Institute of Technology, made a comprehensive ergonomic study of different types of computer jobs and published their characteristics in his highly regarded textbook, *Ergonomics in Computerized Offices*. Most of the descriptions of the various types of computer jobs outlined below are based on those findings.

Data entry involves inputting mainly numbers at high speeds while reading them off documents. Typically, you key in numbers with your right hand, while manipulating documents with the left hand. The documents are usually placed either flat on the desk or upright on a document holder. Your eyes are fixed on the documents while your fingers work like little mechanical levers, tapping on the keys. The speed at which the fingers work is incredibly high—as high as five key strokes per second! That is, the fingers move down then up back to their original position about five times per second. Adding-machine operation is one type of data-entry job. Studies have shown that about 81% of data-entry workers spend more than six hours per day at the keyboard.

Data entry is perhaps the most physically strenuous computer job. In one 1990 survey of 514 data-entry operators and clerical employees from 31 data-processing organizations, 64% of the operators reported that they suffered from pain in the neck and shoulders, 52% from pain in the back, 28% from pain in arms and legs, and 12% from swollen muscles and joints. These figures are typical of data-entry jobs in different workplaces.

Data acquisition, or data inquiry, involves bringing information from the computer brain onto the screen for viewing. You spend almost all of your time looking at the screen. A telephone information operator is a typical data-acquisition job.

Conversational, or **interactive**, terminal jobs involve operations like payment transactions at banks and airline flight reservations. You enter data via the keyboard and observe the feedback on the screen in order to make further decisions. You may then make changes or continue entering more information. Typically, both hands operate the keyboard. Your head and eyes are usually directed at the computer screen about half the time and the documents the other half. If one or both of these visual displays are not positioned directly in front of your eyes, you may be forced to twist your neck, rotate your eyes, or both. The documents often lie flat on a desk. Unlike data-entry jobs, the quality of information is more complex than simple numbers and words so mental activity may be more significant. This mental activity helps to ward off boredom, and speed of entry is not critical so keystroke speed is relatively low. In some types of conversational jobs, such as airline reservations, information is copied from documents to the computer. All information is displayed on the screen. About 73% of conversational terminal workers spend more than six hours a day at the keyboard.

Computer-aided design and manufacturing (CAD-CAM) operations are the computerized versions of the old engineering drawing board operations. In today's hi-tech world, this job involves a lot of mechanical design, printed circuit board design, and electrical schematics design. A decade ago the engineers used a pen and a drawing tablet about 40% of the time and a keyboard about 20% of the time. Today their eyes are fixed on the screen about 60% of the time for examining the design graphics. CAD-CAM operation is not as intensive or stressful as data entry. Workers get more opportunity to move various parts of their bodies, reducing the strain of maintaining constrained posture. The mouse and trackball have replaced the drawing tablet. They are more versatile than the drawing tablet, but many workers think that they have increased the intensity and, hence, the stressfulness of CAD-CAM work—just as the desktop computer did for word processing when it replaced the typewriter. Engineers working at computers also tend to maintain fixed postures for longer stretches of time than those who worked with a pencil and paper.

Word processing involves mainly creating text. This may involve copying information from documents or dictating machines, recalling text, or creating design layouts. It involves a lot of the typing work performed by secretaries in the past, but it is now being done with a computer instead of a typewriter. The use of the mouse and Windows-type software have simplified and sped up the work but have not necessarily reduced the amount of time workers spend sitting in front of the computer. Word processing is much more intense than traditional typing. Typically, both hands operate the keyboard or one hand operates a mouse, and the speed of work is fairly high. The eyes are usually directed at the documents while typing. About 65% of typists spend more than six hours per day at the keyboard.

In **secretarial, or administrative assistant, jobs** there is greater variation in tasks. The keyboard is not used as often as in other jobs, and there are adequate opportunities for moving various parts of the body, such as using the phone, filing papers, and meeting with coworkers.

POINTS TO REMEMBER

- There are many different types of computer jobs, and they cause different types of stresses on your body at different levels.
- Not all computer jobs cause persistent aches and pains.
- The most stress-inducing jobs are those in which you perform excessive keying or reading (from screen or documents) or remain in one posture for prolonged periods of time.
- Data entry is probably the most stressful computer job, which places excessive stress on the hands and wrists.

CHAPTER 6

ORIGINS OF COMPUTER-RELATED PAIN

The perception of aches and pains is an important function of your body. Pain from working muscles is a signal from your brain saying that a part of your body is under excessive strain. It is a defense mechanism—a kind of safety alarm from the brain telling you to slow down or stop working. The message is that your body needs rest to recuperate and prevent impending damage to its structures.

The aches and pains a computer worker feels at the end of the workday are not unlike those felt by workers performing many other types of physical work. The exact causes may be different, but the effects on the body are similar. If you perform a lot of computer work, your wrists, lower back, the back of your neck, and shoulders are likely to be overwhelmed by sustained strain. Under certain conditions of work, your legs are also stressed. Wrist problems due to repetitive motions will be discussed in chapter 12. Shoulder, neck, back, and leg problems are due to muscles maintaining a state of static contraction for prolonged periods while sitting. This effect is called **static overloading** or **postural overloading** of muscles. Our experiences at computer terminals should convince us that we do not necessarily have to handle heavy loads to suffer from musculoskeletal strains. Simply sitting and working can lead to such mechanical strains in the body. But how exactly does this happen? The answer to this question is found in the way our bodies react to resist the pull of gravity.

Gravity and Muscular Work

Gravity always affects us. It affects subtle functions related to our growth, such as the flow of blood and the shaping of the body, the methods and the efficiency of our motions in such activities as walking, running, jumping, or climbing stairs, and the way we maintain our body posture. In all of these activities, the muscles of the body must work in order to counteract the effect of gravity. In traditional office work, before the advent of the desktop computer, gravity was never considered a source of problems because muscular activities did not exceed the limits of muscular capacities. But today the prolonged sitting in the computer workplace has cast gravity in a new role. Gravity is like a demon in the computer workplace, pulling and tugging at our bodies with forces that are impossible to eliminate by ordinary means. But if we understand how gravity imposes these forces on our bodies, we can minimize them and save our bodies unnecessary wear and tear and consequent musculoskeletal problems. To understand how the computer workplace causes musculoskeletal problems we must start with an understanding of the musculoskeletal system, which is described in chapter 7.

Body Posture and Muscular Work

The configuration of the various segments of your body at any particular time is called **posture**. In maintaining a particular posture, no part of the body moves and no external object is moved. However, your body is constantly being pulled downward by gravity, and certain muscles must constantly contract to oppose the pulling force. Your head, arms, trunk, and legs are constantly being pulled downward, and if it wasn't for the work of the muscles in these body segments, your body would slump to the floor. The reason you cannot get a sleeping or unconscious person to sit or assume any stable posture other than lying down is because his muscles won't contract enough. While sitting and working at the computer workstation, many muscle systems in the body work hard against gravity just to stabilize the body on the chair. Those in the neck, back, and arms work the hardest.

Gravity's pull on a particular part of the body depends only on its weight and is fairly constant over a short period of time. But gravity's pull on one part of the body has an effect on other parts, and this effect varies from posture to posture. For example, gravity's pull on your trunk is the same whether your upper body is upright or bent forward. In other words, your trunk weighs the same in any posture. But when you bend forward there is greater strain on the lower back than when you are upright. The muscles in your lower back must work harder to maintain the bent posture. Sit and bend forward and you will experience this effect as a tightness in your lower back.

Other examples of common, stressful computer-working postures are bending your head and neck sharply forward or sideways and holding your arm outstretched. The former requires stronger neck muscle contraction than when the head is upright, and the latter requires stronger shoulder muscle contraction than when holding the upper arm at the side of the body. What is true for the back and neck is also true for other body segments. The extra work required to maintain certain positions is what leads to aches and pains. Whenever we shift in our seat to ease the aches in certain muscles and joints, we are essentially fighting against the force of gravity. Any attempt to minimize extra muscular effort associated with a body posture is a fight against gravity.

Reducing Muscular Work

Trying to prevent aches and pains while sitting in front of a computer is a challenge to minimize the effect of gravity on your body. We cannot actually change the force of gravity, but we can change the way it acts on us. We can do this by changing our sitting postures. The way we sit at work can determine whether we have severe back pains, mild discomfort, or no problem at all at the end of the workday.

Throughout history, the fight against gravity has been centered on moving heavy objects. Powerful machines were built to perform tasks that were too strenuous for our muscles. A small amount of muscle force was input into a machine, which then produced an output force many times greater. Basically, machines amplify human strength. The crane is an example of such a successful machine. A single pull or push on a lever is translated into machine force that can lift thousands of pounds.

The fight against gravity has now entered the office, but we are hardly aware of it. This fight is aimed not at amplifying muscular strength but at preventing situations in which muscles are forced to work beyond their capacities. The basic technique is to control body posture. This is the principle on which the design of ergonomic furniture is based. Almost all of the features of this furniture are designed to reduce muscular effort. Evaluating your chair, your desk, and the arrangement of other equipment in your workplace will help reduce muscular effort as well. Here are some of the crucial questions you will need to answer:

- Am I sitting in a stressful posture?
- What is imposing that posture on my body—furniture design, workplace arrangement, job characteristics, or a combination?
- How can I avoid this stressful posture?

In order to answer these questions, you must understand the system inside the body that is designed to oppose gravity, which is the arrangement of bones and muscles, or the **musculoskeletal system**.

Static Versus Dynamic Work

One of the most obvious things you will notice about someone working at a computer is that the fingers move continually while the other parts of the body remain still. In this state, the fingers are said to be working **dynamically**, while the other parts of the body work **statically**. Dynamic muscular work involves motion. The muscles contract and relax alternately, moving the body segments they are attached to. Static muscular work involves no motion in the muscles or body segments attached to them. The muscles merely contract and maintain their tension without shortening or lengthening. Almost all work involves a combination of static and dynamic muscular contractions. Dynamic contraction may be concentrated in one part of the body and static contraction in another part, as in computer work.

Many jobs were dynamic in nature until machines were developed to replace muscle power and duplicate manual motions, increasing static work. Although the machines performed the work, humans controlled them. This often required keeping the body, or part of it, still while concentrating intensely. Computer work is similar because it requires you to press specific keys at specific times and at a certain rate, and keep your eyes fixed on a document or the screen. This type of job is performed best when you sit still and keep all parts of the body, except the fingers, motionless.

In addition to back muscles, many other sets of muscles contract and maintain their tension during computer work to hold your body in a particular working posture. Neck muscles must hold the head still to permit you to read from the screen or document, trunk muscles must hold the upper body in a particular position to bring your head and arms at the correct distance from the computer, arm and shoulder muscles must hold the hand in position over the keyboard, and leg muscles must stabilize the lower body to maintain balance on the seat. As you can see, computer work is much more strenuous than it might appear. The muscles that perform static work simply do not get the credit they deserve for their efforts. Similarly, you may not get the full credit you deserve for enduring the stress of this kind of work.

Keeping your body still is considered strenuous because static contraction encourages the accumulation of poisons in the muscles, leading quickly to fatigue. But exactly how does this happen and how can we prevent it from happening?

How Lactic Acid Poison Accumulates in Muscles

The poison in a muscle is mainly a waste product, called lactic acid, formed when food substances in the body are broken down to release energy. A muscle requires energy to contract. This comes from the foods you eat. After digestion in the stomach and small intestines, food material diffuses into the tiny capillary blood vessels surrounding them. From there the food is taken to your muscles where it is broken down with the aid of oxygen (oxidized) in a complex biochemical process that releases energy. This process is called **aerobic metabolism.**

The oxygen comes from the air you breathe. Your lungs are lined with tiny blood vessels. Oxygen passes from your lungs into the red blood cells that take it to muscle cells. Carbon dioxide, water, and heat are produced as wastes and are excreted from your body. This process is similar to the energy production in a car engine where gasoline is like the food and the engine is like the muscles. Gasoline is fed into the engine where it combines with oxygen from the air that has been sucked into the engine. A chemical reaction in the engine then generates the energy that moves gears and levers. Carbon monoxide, carbon dioxide, and water vapor are produced as waste products and flow out of the engine via the exhaust.

As your muscles work harder, your heart beats faster to increase blood flow and take more food and oxygen to the muscles. But there are times when your blood cannot supply enough oxygen or food substances to produce energy at the rate demanded by the muscles. During dynamic work, this occurs when your muscles are contracting and relaxing too fast or too forcefully, even though blood flow to the muscles can increase as much as twenty times. During static work, insufficient oxygen supply occurs because blood flow through the statically contracting muscles is inhibited. In both situations, a back-up energy-producing mechanism, which does not require oxygen, is activated. It is called **anaerobic metabolism,** but it comes with a cost. It produces lactic acid as a waste product.

Naturally, wastes produced in any part of the body must be removed or they will accumulate inside the muscle cells and prevent them from functioning properly. If lactic acid is removed as fast as it is produced, no harm comes to the muscles. If it is not, it builds up inside the muscles and poisons them, inhibiting the muscles' ability to contract properly and gradually weakening them. This is what is called **muscular fatigue**. The symptoms of fatigue are, first, a feeling of tiredness, then mild aches, and finally pain. This is the way you probably feel at work as the day progresses.

Protection against Muscle Poisoning

The body has a natural protection from lactic acid poisoning. Metabolic wastes are withdrawn from the muscles into the blood as it flows through the muscles. The wastes are then taken to the kidneys, lungs, and skin, where they are excreted as urine, carbon dioxide, and sweat. In this way, the blood cleanses the muscles, metabolism continues uninterrupted, and you are spared from fatigue. But this cleansing process requires the blood to flow freely through the muscles. Static work obstructs free blood flow and upsets this cleansing mechanism.

How Static Work Slows Down Your Blood Flow

The flow of blood through the arteries, capillaries, and veins is maintained by the pumping action of the heart and the rhythmic contraction of the blood vessels. During dynamic work, the muscles contract and relax alternately. Contraction squeezes the blood vessels within the muscles and forces blood along its path. Relaxation allows the blood vessels to expand to normal size and pull fresh blood into the muscles. This squeezing and relaxation creates a sort of pumping mechanism that facilitates the free flow of blood. It is considered a secondary pump in the cardiovascular system. The heart is, of course, the primary pump. As long as dynamic activity is not too strenuous, this two-pump mechanism maintains adequate flow of blood through the muscles and sustains work without fatigue. Enough energy-rich food is brought to the muscles; enough oxygen, trapped in the red blood cells, is brought to the muscles to oxidize foods; and waste products, including any amount of lactic acid from metabolism, are flushed out of the muscles.

During sustained static work, such as when you sit still at a computer, the effect is different. The blood vessels within the muscles are squeezed for as long as muscle contraction is maintained without relaxation. A squeezed blood vessel has a more narrow passage, like a squeezed garden hose, and blood flow is reduced. The stronger the contraction, the greater the reduction in blood flow. When muscle contraction is about 25% of its maximum strength, blood flow restriction begins. When it is about 70%, blood flow through the muscle is completely cut off. Sitting still and working at the computer automatically produces some restriction in blood flow. When you are in a constrained posture, muscles in the neck or trunk are more strongly contracted, and blood flow is further restricted.

When blood flow is restricted, the muscles become robbed of adequate nutrients and oxygen and are forced to resort to anaerobic metabolism to compensate for the deficit. Lactic acid waste is then generated, but, because of reduced blood flow, it cannot be removed as fast as it is formed. It accumulates in the muscles and poisons them. This is when you feel the familiar aches and pains. This is the point at which you usually get up and walk about or rub these aching muscles. You are then enhancing normal blood flow and reactivating the cleansing mechanism.

Preventing Lactic Acid Accumulation

From the preceding discussion it is clear that excessive static muscular work should be avoided because it promotes the buildup of lactic acid to harmful levels in muscles. Since computer work cannot be done properly unless the body (except the fingers) is kept still, some lactic acid accumulation is inevitable. Getting up and walking during break periods provides opportunities for the muscles to work dynamically. So do muscle stretching exercises. Free blood flow through the muscles can then be resumed, and lactic acid will steadily be removed. The muscles will also begin to receive more nutrients and oxygen and produce less lactic acid.

Remember that getting up from your seat produces other benefits. It relieves mechanical tension in muscles, tendons, and ligaments and mechanical pressure in joints. Your body can then repair itself from wear and tear and delay the onset of further fatigue. Getting up also takes some of the weight of your upper body off the **lumbar spine** and seat bones and transfers it to the legs, which have the strongest muscles, reducing pressure in the lumbar spine and buttocks. Chapter 16 describes the methods of combating fatigue in greater detail.

POINTS TO REMEMBER

- The sensation of pain is a signal from your brain telling you that the part of your body experiencing the pain has been overworked.
- Gravity is constantly pulling your body downward, and muscular contraction opposes that pull.
- Muscles work harder in certain postures than in others. Back muscles work 35% harder when sitting upright without a backrest compared to standing and 95% harder when leaning forward.
- The energy for muscular contraction comes from metabolism on the foods you eat. Metabolism occurs in muscles.
- Muscular work above a certain level produces poisonous wastes in the muscles.
- Dynamic contraction of muscles, as in walking, swimming, etc., helps the blood flush lactic acid out of muscles.
- Static contraction, such as occurs in sitting, squeezes blood vessels, slows down the flow of blood, and causes lactic acid to accumulate in the muscles. Back, neck, and shoulder muscles contract statically when sitting and working at a computer.
- Muscular aches and pains are felt when lactic acid accumulates in muscles.
- You can avoid aches and pains by not sitting still for too long and by avoiding postures that require strong muscular contractions.
- Rubbing and flexing aching muscles helps prevent accumulation of poisons in them.

CHAPTER 7

RESISTING STRESS ON MUSCLES AND JOINTS

Every large complex structure requires some sort of framework on which to maintain its form, shape, and strength. Buildings, for example, are built around a network of steel rods or some other design. But inanimate things are not the only structures that require a framework. Many animals have them. In animals they are called a skeleton. In human beings, the skeleton is a complex arrangement of 206 rigid bones of various shapes and sizes. At the ends of some of these bones is a less rigid substance called cartilage. Bone is composed of calcium and is brittle, and cartilage is composed of compact connective tissue and is not brittle. It protects the end surfaces of the bones from wear and tear as they slide over, compress, or impact one another. In your spine, this kind of wear and tear is associated with such work as driving a tractor or repetitive lifting. It is suspected that computer work, under certain conditions, can also cause it. Sitting in constrained postures for prolonged periods, for example, can create so much pressure in the spine that certain vertebrae may gradually wear and damage the less durable discs between them.

Sitting Posture and Movement

In order for us to move, our skeletons must be able to bend at certain points, and forces must be applied to these points to achieve the bending motions. Bending is made possible by joints. Forces are produced by muscles attached to the bones across these movable joints. These are nature's engineering designs that make motion and change of posture possible.

Sitting requires a change in the geometrical configuration of the body. This is brought about by the movement of certain body segments around joints. For example, if you want to move from a standing to a sitting posture, you must bend at the knees and hips. To use a keyboard while sitting, you must rotate your forearm upward until it is approximately horizontal. These geometric changes can build up considerable strain when maintained for prolonged periods. To understand this, you must understand how the structure of the skeleton and muscles facilitates these motions and postures.

Joints—Movement and Posture

A joint is a junction where the ends or edges of adjacent bones meet and bind in a special way. It allows the bones to move relative to one another and permits the body to move or change posture. Some joints, like the shoulder and wrist, have great freedom of movement. Others, like the elbow and knee, are more restricted, and those in the skull are immovable. There are almost two dozen joints along your spine. Each has a limited range of motion, but collectively they let you bend your trunk sharply in many directions.

The mobility and flexibility that movable joints impart do not come without a cost. The cartilage surfaces forming the joint may wear and tear as they move over one another under mechanical pressure. This is common in the joints of your spine, especially when performing jobs that impart heavy loads on the upper body. Heavy lifting is obviously such a job, but simply sitting and bending forward may also create considerable pressure in your spine.

In addition, muscles and tendons increase the pressures in joints considerably when they contract.

Muscles and Tendons

Muscles are formed from specialized cells that have the ability to contract and generate forces. There are three main types of muscles in your body: **smooth, cardiac,** and **skeletal**. Smooth muscles are found in certain organs, such as the bladder, the digestive tract, and blood vessels. They contract automatically, and the forces they generate serve to propel substances such as urine, food, and blood. Cardiac muscles are the muscles of the heart. They also contract automatically, and their regular, periodic contraction propels blood out of the heart.

The third kind of muscles are those attached to your bones by tendons called skeletal muscles. They are also called **striated muscles**. These muscles are responsible for movement and maintaining posture. They are the ones associated with the aches and pains from physical overexertion in computer work. There are more than 430 skeletal muscles in your body, and they account for 40% to 45% of your body weight. They occur in pairs on opposite sides of your body. The most vigorous movements of your body are produced by fewer than eighty pairs. Throughout the rest of this book, any mention of the word **muscle** will refer only to the skeletal muscles.

When a muscle contracts, it pulls on the bones it is attached to, causing the particular body segments to move or to maintain a fixed position against the pull of gravity. For example, if you want to hold your trunk in a bent forward position, certain back muscles contract just strongly enough to nullify the effect of gravity's downward pull. In other words, when you sit bent forward, your trunk muscles are in a state of constant contraction. If the contraction force is greater than the force of gravity's pull, your trunk moves upward; and if it is weaker, your trunk moves downward. If the two forces are equal, you remain still.

The ends of a muscle are rarely attached directly to bones. Instead, most muscles merge into cord-like structures called tendons, which attach to the surfaces of bones. Tendons are made of strong connective tissue. Their job is to transmit muscle forces to the bones. It is not difficult to understand how a powerful muscular force can tear either a muscle or its tendons in the same way that a strong pull can tear a rope. However, it is often difficult to accept that forces that are relatively weak can also tear muscles or tendons. But if they are repeated often enough, or if they are maintained for prolonged periods of time without adequate rest for recuperation, they do. In such a case, the tearing is gradual and accumulates over time until pain is felt. This is what happens when you sit and work in certain postures at a computer workstation.

An example of repetitive forces in computer work is the rapid keying on the keyboard. This activity is thought to be responsible for the gradual wear and tear of muscle tendons in the wrists and hands. Sitting at your computer is an example of muscle force being maintained for prolonged periods in computer work. The muscles and tendons in the trunk, back, and legs are tensed for as long as you sit and work. They are relaxed only when you take a break.

Stabilizing the Joints

Ligaments, like tendons, are cord-like structures made of compact connective tissues. However, their function is not to move segments of the body but to prevent excessive movements. They connect one bone to another across a joint and are responsible for keeping the joint intact. They prevent the joints from dislocations or subluxations (moving out of alignment) when they are subjected to great forces. Such forces can come from outside the body, as during a car wreck, or from inside the body from muscular contractions, as in sitting and working in front of a computer with your neck sharply bent.

Many people work with their necks bent forward. When the angle is sharp, the ligaments in the neck tense to oppose this bending and prevent the neck bones from moving out of position. But working for hours every day with your neck bent can eventually stretch and weaken the ligaments and joints. Weakened ligaments reduce a joint's resistance to compression and other types of mechanical pressure. If the neck joints and their ligaments have already been weakened, as from diseases or whiplash injuries, working with a sharply bent neck can further weaken them and produce subluxations more easily. In extreme cases, a disc herniation in the neck may occur.

The Spine

The human spine is the important part of the skeleton that protects the spinal cord. It warrants a separate section in this book because of its importance in understanding stress from sitting and because of its peculiar anatomical structure. Understanding the structure of the spine can help you understand not only how pains arise in the back and neck from sitting and working but also how you can prevent those pains. You will be able to understand how a sitting posture can cause a change in the shape of your spine and how pressure is built up from this change. In turn, this will allow you to determine what kind of chairs are best for you, how to sit on them, and how to arrange your workplace to prevent pains in your back and neck. You don't always have to wait on your employers to assign an expert to deal with problems, and you don't have to rely on an ergonomist, a physician, or anyone else to prevent work-related pains from occurring in your body. You can do a lot for yourself.

The **spine**, also called the **backbone** or **vertebral column**, is that part of the skeleton that extends from the base of the skull to the pelvis. The spine consists of a chain of thirty-three bones, arranged one on top of the other to form a column (see Figure 1). Each bone is called a **vertebra** (**vertebrae** is the plural form). Only the top twenty-four, which are above the pelvis, are movable as individual bones. The bottom nine are fused into two larger bones, called the **sacrum** and **coccyx**. So you could say that the spine consists of twenty-six bones.

The vertebrae in the spine form four distinct regions. The top seven are in the neck and form the **cervical region**. Going down the spine, the next twelve are in the chest or **thorax,** and the next five are in the lower back or **lumbar region**. The next five are fused to form the sacrum, which is wedged between the two halves of the pelvis, and the last four are fused to form the tail-like coccyx. The coccyx gets its name because of its resemblance to a cuckoo's beak. People who believe in the evolutionary theory of human beings consider the coccyx to be the vestiges of a tail from our apelike ancestors.

Discs—Cushions in the Spine

Each vertebra is separated from its neighbor, above or below, by a special disc-shaped sac called a **vertebral disc** or simply a **disc.** These discs are the special joints of the spine that account for about a quarter of its length. A disc isn't as hard and brittle as bone. It has a tough, fibrous outer covering called the **annulus fibrosus**, and a pulpy inner core called the **nucleus pulposus.**

Discs are compressible like water beds. They act as cushions for the mechanical shocks you get when walking, running, or jumping, and for the sustained pressure you experience while sitting and bending forward. They also help to hold the vertebral bones together and provide flexibility, strength, and protection to the spine. It is mainly because of these discs that you can bend so easily at the waist and neck. When you grow older, your spinal discs may degenerate and become thinner through loss of fluid, causing you to "grow downwards." This also causes your spine to become weaker and less flexible.

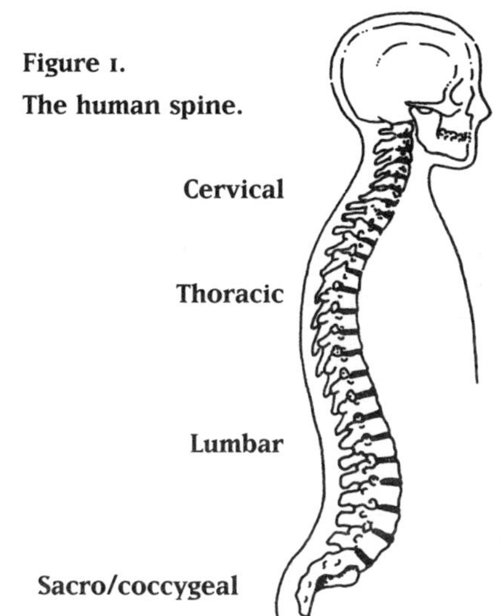

Figure 1. The human spine.

Despite their resilience, discs can degenerate from excessive or sustained load on the spine. Such a damaging load can occur from sitting and working in constrained postures. Sustained pressure on the spine can cause the cartilage endplate surfaces of the vertebrae to degenerate, and because these surfaces are in contact with the discs, the annulus fibrosus wall of the discs can be torn. Eventually the nucleus pulposus core may extrude through the torn wall, causing a slipped (or herniated) disc.

If the vertebrae of the spine were not separated by a compressible structure like a disc, their surfaces would erode gradually. The weight of your upper body would erode them like the soles of your shoes. Vertebral bone surfaces would also be crushed easily without the protection of discs, when motion or shocks occur from actions like brisk walking or jumping. Wherever two vertebral bone surfaces meet, there would be rapid wearing away of the surfaces and consequent degeneration of the spine. Of course, human beings could not have survived under these conditions. The discs protect the surfaces of the vertebrae from this kind of wear and tear.

The way the spinal discs cushion excessive pressure is similar to the way a water bed cushions your body. When you lie down on a water bed, your weight presses down on one part while the water presses outward on other parts. In the same way, your upper-body weight presses on spinal discs, and the jelly-like nucleus pulposus presses outward on the side wall. The wall is, therefore, always under pressure. But this wall, in a healthy disc, is strong enough to withstand great pressures. A weakened or torn disc wall, however, can be punctured by the internal pressure from the nucleus pulposus, leading to a slipped disc.

Shock absorption involves alternate compression and relaxation of the discs. This creates a kind of pumping action that also helps the discs feed themselves. Discs are living tissues but, unlike muscles, are not perfused with blood vessels that supply nutrients dissolved in the blood. Tiny blood vessels called **capillaries** run just outside the body of a disc, and the pumping action in the discs from body motions draws food out of these capillaries into the discs by a process called **diffusion**. If the spine is kept motionless, as when sitting still and working at a computer, the pumping action stops and the exchange of food and other substances between the discs and capillary blood vessels slows down.

When sitting and working in front of a copmuter, there is a tendency to remain still for prolonged periods, especially if you are concentrating intensely or trying to maintain a high speed of keying. The discs are then under sustained pressure by upper-body weight. They do not get the opportunity to relax. Diffusion is therefore inhibited, and the discs are robbed of proper nutrition—one more reason computer work can be hazardous to your health. So when sitting at work, it is a good idea to do some bending exercises to enhance the pumping mechanism. Bend your spine slightly in many directions to activate blood flow and the food diffusion process. Getting up from your chair and walking around is often as good as any exercise. People who occasionally stretch their backs to relieve muscular fatigue inadvertently activate this mechanism, to their benefit. Rubbing your back with your hands also helps, as does adjusting your chair to modify your posture.

Concentrated Strain in the Spine

Work-related aches and pains in the spine are concentrated in the neck and lower back areas. Lower back problems are more common among people who perform heavy or repetitive lifting tasks, but they also occur during computer work. Neck problems are even more common among those who sit and work. It is estimated that at least 60% of the population suffers from some sort of lower back pain, and in about 60% of these sufferers the pain is due to overexertion.

Neck strain has always been common among sewing machinists, teleprinter operators, seamstresses, and microscopists. This is because their work requires looking at things intensely and fixing their heads in one position for prolonged periods. They typically bend their necks forward to bring their eyes close to the workpiece. Office work, however, was never characterized by widespread neck fatigue problems until the desktop computer invaded the office. Computer jobs require intense visual work with little movement of the head for prolonged periods. So it is no surprise that neck strain is common among computer operators. Studies have shown that over 50% of keyboard users suffer from aches and pains in their necks and shoulders.

Because strain in the spine has been concentrated in the neck and lumbar regions, it is tempting to believe that these regions are weak. But in reality, they are extremely strong. The problem is that they are simply overworked. Nature did not design these areas of the spine to sustain the great mechanical stress that modern work imposes on them. For example, your body was not designed to sit for long periods of time. Activities that place excessive stress on these areas of the body are health risks. Sitting and working at a poorly designed computer workplace is one such health risk, and it is widespread today.

Overexertion at the neck and lower back is caused by one or more of the following reactions in the body:

- Excessive mechanical pressure in discs
- Excessive pull in muscles and tendons
- Excessive pull in ligaments

These strains arise from the weight of the upper body due to gravity's pull and are intensified by unnatural and unaccustomed postures.

Mechanical Pressure in Discs

Approximately 60% of the weight of your body rests on the base of your spine. If you weigh 140 pounds (the average body weight for the American adult female), 70 pounds are resting on the base of your spine.

Any particular part of the spine, such as a vertebral bone or disc, bears the weight of the body above it. For example, the topmost vertebra of the spine, called the **atlas**, supports the weight of the head alone; the vertebrae in the midregion of the spine support the weight of the chest, upper arms, neck, and head; and the lumbar (lower back) vertebrae support the weight of the whole upper body. The weight of the arms also presses down on the spine, because the arms hang from the shoulder. So about two-thirds of your body weight exerts pressure on the lumbar spine whether you're sitting or standing. When you sit at work, the effects of the weight on the spine do not diminish unless you lean against a backrest or on your desk with your arms. Unfortunately, in computer work there is a limit to how far back you can lean on your backrest, and in most of these jobs you rarely have the opportunity to lean on your arms.

If you want to conceptualize how the weight of your arms presses down on your spine, here is an experiment that will help.

Try holding a fairly heavy object in each hand and notice how the muscles in your back contract. If the weight is heavy enough, you will feel the increased pressure in your spine. The effect of the weight is to make your arms heavier. The same effect occurs when you hold an object with your arms in front of you.

The vertebral bones and the discs in the lower part of your spine are thicker and stronger than those in the upper part because they must support more weight. If the vertebrae lower down the spine were as small as those at the top, you would be suffering from fractured back bones constantly. Your spine would have been crushed from impacts due to activities like jumping or sudden lifting of heavy objects.

Spinal Pressure

The way the upper-body weight and muscles impart pressure on the spine can be explained by an analogy with a ladder. Suppose you can stand a ladder perfectly upright on the ground. Its weight will then compress the ground. If you tilt it, it will fall. You can prevent the fall by pulling downward on a rope tied to it near the top. The force pressing down on the ground will then be a combination of part of the ladder's weight and part of the pull force on the rope. Combined, these forces are much greater than the force of the upright ladder's weight alone. The strength of the combined force depends on the angle of inclination of both the ladder and the rope. There is also a sliding force at the bottom of the ladder, which is stronger when the ladder is sharply inclined and weaker when the ladder is more upright.

Now consider yourself sitting with your trunk upright and your spine in its normal curvature. This is analogous to the ladder standing upright (see Figure 2). The compressive force on your spinal discs will be due only to the weight of your upper body. When you bend your trunk forward, the muscles and tendons attached to the spine will pull on it to prevent you from toppling over, in the same way that the rope pulls on the inclined ladder. This pull force will increase the pressure in the spine. The discs lower down the spine will be more affected than those higher up. This is partly because the lower discs bear more body weight than

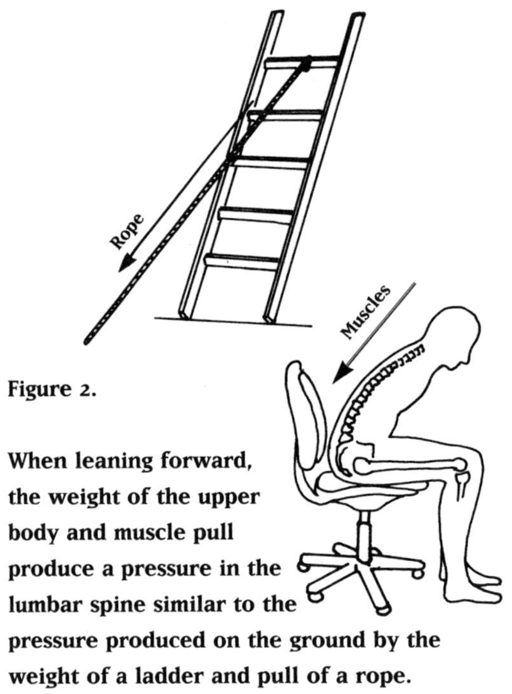

Figure 2.

When leaning forward, the weight of the upper body and muscle pull produce a pressure in the lumbar spine similar to the pressure produced on the ground by the weight of a ladder and pull of a rope.

the higher ones and partly because the upper body pivots on the lowest lumbar disc. This pivot, the disc between the lumbar and sacral vertebrae, is called the **lumbosacral disc**. If your neck is bent forward, similar changes occur in its vertebrae and discs. The muscles at the back of the neck will pull harder to maintain the head in position, and the pull will compress the discs. The pivot for the head and neck is the disc at the base of the neck.

Pressure from prolonged bending of the spine can gradually wear the cartilage surface (endplate) that separates a vertebra from the adjacent disc. If this surface becomes worn enough, it can gradually tear the disc wall (annulus fibrosus). The weakened wall will then be unable to withstand the pressure from inside the disc. This kind of degeneration may cause the disc to lose some of its water and become thinner, more brittle, and weaker. Activities that were once easy on the body, such as bending or carrying light weights, can then become painful on the neck or lumbar spine. In extreme cases, as in sharp bending for prolonged periods or heavy lifting, the nucleus pulposus inside the disc may extrude through the weakened and ruptured disc wall (disc herniation). This usually happens at the rear of the disc, causing the nucleus to press against the spinal cord. Even before herniation occurs, however, the weakened and bulging disc can press on the ligaments at the rear and gradually weaken them. The pain from weakened ligaments is always bothersome, but the pain from pressure on the spinal cord is unbearable and crippling.

The sharper you bend your trunk or neck, up to a point, the

greater the pull in the back or neck muscles and the greater the pressure on the spinal discs. To give you an idea of the great contribution of back muscle pull on spine compression, consider the following approximate figures. Let's say you weigh 140 pounds. If you sit and bend forward only ten degrees, the total compression in your lumbar spine would be 275 pounds—138 from muscle pull and 137 from your upper-body weight. For a thirty-five degree bend, the compression would be 530 pounds—455 from muscle pull and 75 from the upper-body weight. From these figures, you can see that back-muscle pull is responsible for most of the increase in spinal compression when you bend sharply.

A muscle pull force of 455 pounds is like the tension in a rope with a weight of 455 pounds tied at its end and suspended in space, and a compression of 530 pounds is like a weight of 530 pounds resting on the spine. Sitting bent forward thirty-five degrees is, therefore, an incredible strain to bear for even an hour. Such a posture, maintained for a few hours a day over several months or years, will most likely lead to chronic back aches and pains. (A compression of about 2,000 pounds within the spine can crush it.)

Ligament Strain

The ligaments that bind the vertebrae can also be strained when your trunk or neck is bent in any direction. Remember that the function of ligaments is to bind and maintain bones together at a joint. Bending towards one side of the body automatically tenses the ligaments on the opposite side as they try to resist dislocations and maintain the bones in proper alignment. The lumbar and neck regions of the spine are most affected since they are the most flexible parts of the spine. Some of the pains that you feel in the lumbar and neck regions from prolonged sitting are due to strained ligaments. The more sharply bent your trunk is when sitting, the greater the pulling and straining of the ligaments.

Neck Strain

The weight of your head and neck is about 8% of your body weight. For an American adult female of average weight, this is about eleven pounds, and for a two-hundred-pound person, it is sixteen pounds. In other words, the weight of the head and neck is about the same as the weight of a bowling ball. Imagine the strain in the neck muscles when you are working at the computer with a sharply bent neck. Engineering calculations on the human body indicate that, for a ten degree forward bend of the neck in an average-weight female, the pull in the muscles is about five pounds. For a thirty-five degree bend, it is twenty pounds. As you can see, working with the neck bent thirty-five degrees imposes tremendous strain in the neck muscles and spinal discs.

You can get a good idea of the muscular work required to maintain your neck bent by performing the following experiment. Insert a broomstick (or some other rod) into one of the holes of a bowling ball. With your hand placed several inches below the ball, hold the stick and ball upright. Now, allow the stick to incline forward like a bent neck and note the increase in the muscular strain in your arms as you try to maintain the inclined position. This will give you an idea of the strain in the neck muscles required to stabilize and maintain your head in a bent position. The way in which the back muscles increase their contraction to maintain an inclined trunk follows the same principle.

If you replace the ball with a heavier one, or if you increase the angle of inclination of the stick, the stabilizing force increases. In other words, the heavier your head and neck or the more sharply bent they are, the greater the pull in muscles and the compression in the spine. There is a limit for muscle pull. When a joint is bent beyond a certain point, the contraction of the muscles and tendons can decline, leaving the strain to be taken over by the ligaments spanning that joint. Sharp bending of joints tends to dislocate them, and it is the role of the ligaments to oppose and prevent such dislocations. For the trunk, this occurs at about sixty degrees flexion.

Twisting Force in Your Spine

A common sight in the computer workplace is a person sitting with his trunk or neck rotated, often to the left, and his head looking down at documents on a flat desk. In this case, the upper body exerts a twisting force on the spine. Engineers call this force **torsion**. Torsional strain is concentrated in the lumbar region of the spine when you rotate your trunk and in the neck when you rotate your head. It easily leads to pains if the rotation is sharp or maintained for prolonged periods of time, even if it is not sharp.

The great effect of torsion can be demonstrated by using the analogy of clothes wringing. If you hold the two ends of a wet towel in your hands and wring them, water is squeezed out easily. This is due to the torsional forces acting within the cloth.

Torsion in the spine can wrench the vertebrae of the spine apart. The muscles that create the forces for this twisting and the ligaments that hold the spinal vertebrae together will also be pulled tightly. Since ligaments run from one vertebra to another, when the spine is twisted they contract strongly as they oppose the torsional forces. The muscle tendons are also tensed. For these reasons, you should avoid twisting postures while working at a computer to avoid some of the familiar aches and pains in the neck and lower back you feel by the end of a workday. Trying to avoid glares on the computer screen and reading from documents on the desk are two common work conditions that force people to sit with their trunks and spines twisted.

Obesity and Back Strain

As explained earlier, the heavier you are or the more sharply bent your trunk is when sitting, the greater the back-muscle force necessary to maintain a sitting posture. Does this mean that bigger people feel greater strain than smaller people? Not necessarily. In general, bigger people have bigger and stronger muscles to cope with the greater mechanical stresses imposed by heavier body segments. Problems arise, however, when people become obese, especially if they gain weight quickly.

The strength of a muscle barely increases with sudden weight gain. An increase in body weight of 10%, for example, does not necessarily mean that your strength will increase by about 10%. Nor is there any guarantee that muscle strength will increase at all. So gaining a lot of weight suddenly places undue strain on the back when sitting and bending. The muscles and ligaments are forced to work harder to cope with the extra weight, and the pressure in the spine increases similarly. When bending sharply, the ligaments are also forced to work harder. The big problem with weight gain is that it occurs mostly in front of the spine and creates a downward pull that exerts its effects around the lower back pivot.

Here is an easy experiment you can do to conceptualize the effect of weight gain. Sit on a chair and bend forward. Now allow someone to place a few books, or something else that weighs about twenty pounds, in your hands next to your stomach. Notice the increase in strain in your lower back. This explains why people with large stomachs or breasts are susceptible to lower back strain.

Pregnancy and Back Strain

Women gain a significant amount of weight during pregnancy. According to gynecological data, average weight gain at various stages of pregnancy is as follows:

- At the 20th week—8.6 pounds
- At the 28th week—13.6 pounds
- At the 40th week (end of pregnancy)—30 pounds

This weight gain occurs mainly in the abdomen, in front of the spine, and the body is more strongly pulled downward by the force of gravity. Back muscle tension must, therefore, increase to oppose the extra pull force and maintain posture while sitting and bending forward. As a result, compression in the spine will increase. Small wonder that pregnant women are always burdened by back pains.

Spinal Curvature, Strength, and Strain

The trunk of your body may look straight from the front or back, but from the side it is curved. This curvature is imparted by the shape of the spine. A side view of a model skeleton or an X-ray photograph of your body shows that there are four pronounced curves in the spine, and each defines a different region. If we start from the top of the spine and move down, we will notice a forward curve in the neck region (cervical region), a backward curve in the chest region (thoracic region), a forward curve in the lower back region (lumbar region), and a backward curve in the hip region and below (sacral and coccygeal regions). The forward curves are called **lordosis**, and the backward ones are called **kyphosis**.

The curves in the human spine did not grow accidentally, and they are not an indication of faulty mechanical design of the skeleton. Quite the opposite. They are the result of thousands of years of evolution that have helped humans to maintain an upright posture. In a human fetus, there is only one curve in the spine. It gives the fetus its characteristic curved fetal position. At about the third month after birth, when you begin to hold your head upright, a forward curve develops in your neck region to support your head. Later, when you begin to stand and walk, another forward curve develops in your lumbar area to support your trunk. With these two additional curves, the spine assumes a snake-like appearance with four curves. It maintains this shape throughout your healthy life.

You have probably heard of people with curved spines. In this condition the person has a clearly perceptible sideways curve called **scoliosis**, which is a spinal disease and not the natural curves. Scoliosis gives the trunk a twisted appearance when viewed from the back or front. In healthy people, there is actually a slight, almost imperceptible natural sideways curve in the spine. This is caused by the imbalanced use of the body's muscles. In right-handed people, the curve is toward the right of the body because the muscles on that side are used more frequently and intensely. In left-handed people, the curve is toward the left.

When standing, the curves are normal. But certain body postures can flatten or reverse the natural lordosis in the neck or lower back. This change increases the pressure in that part of the spine. So any body posture that flattens or reverses these curves should be avoided. Sitting and bending forward is an example of such a posture, and it is common in the computer workplace. Scientific experiments have shown that, compared to standing, sitting and leaning forward without the benefit of a backrest almost doubles the pressure in the lumbar discs. Sitting upright without a backrest increases the pressure by about a third.

The spinal curves also enhance the strength of the spine. They help absorb shocks from impacts from such activities as walking, running, jumping, and other motions and protect the spine from fracture or the brain from jarring. In this respect, the spine acts like a spring. Many kinds of impacts, such as those from sudden jerks or jumps, would literally snap the spine if it were straight.

The concept of giving a structure curves to gain strength is found not only in our bodies but also in our creations. Civil engineers and architects utilize this idea in building strong structures, such as bridges and buildings. Even technologically unsophisticated people have used curvatures for imparting strength to their homes. The igloos of the Eskimos and the thatched huts in many rural societies are just two such examples.

Your spinal curves also create a counterbalancing effect on the upper body, preventing any tendency for you to tip over forward or backward. They do this by helping the body maintain its center of gravity over the feet.

Sitting and Back Pains

When standing upright, with the shape of your spine is normal, the surfaces of the spinal discs are in even contact with the vertebrae above and below them (see Figure 3), and mechanical pressure from your upper-body weight is evenly distributed over the disc surface. A change in the curvature of the spine increases the pressure because it redistributes the body weight. One part of the disc surface develops greater pressure than the other parts. This occurs in many activities of daily living, but the increased pressure is seldom harmful. The body was designed to withstand these changes. What matters is how great the pressure increase is and how long it must be sustained.

When sitting and bending the neck forward sharply, your cervical curve flattens or reverses itself into a kyphosis. A similar flattening or reversal of the lumbar spine occurs if your trunk is bent

Figure 3.
The lumbar spine in two different positions of the trunk—erect and forward.

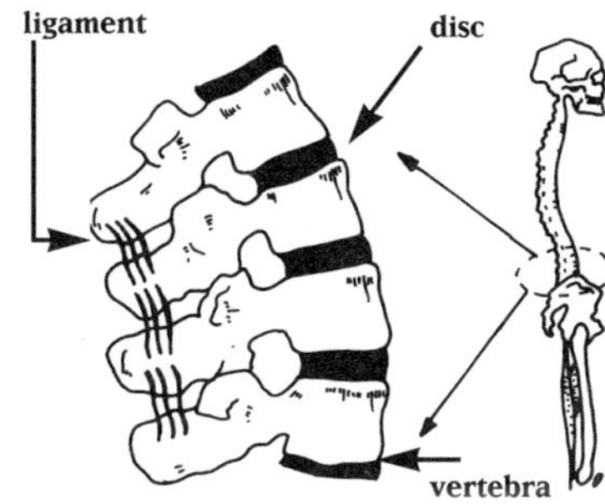

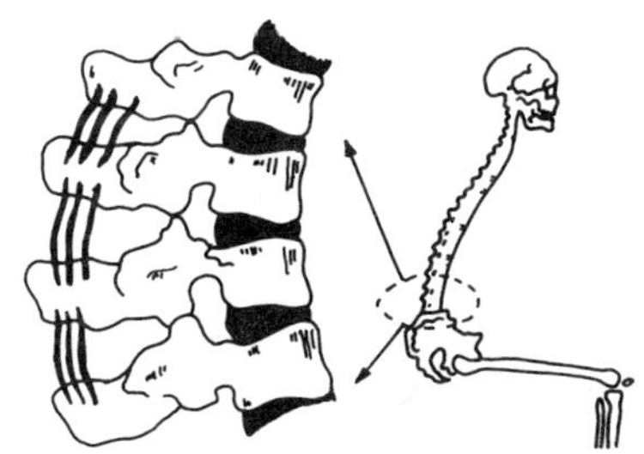

sharply forward. Forward bending shifts the weight above that part of the spine forward and increases the pressure in the front part of the discs. Some of the tightness that you feel in the lumbar spine when sitting and bending forward is due to this increased pressure. If the pressure is great enough, and especially if the bent posture is maintained for a long period, the tightness becomes painful. This happens to many people who work at computers.

An analogy can explain this further. Stand flat on a firm surface and note how the pressure feels on the feet. Your body weight is evenly distributed over the whole surface of your soles. Now gradually raise your heels so that your body weight shifts onto the balls of your feet and note how the pressure there increases. This is similar to what happens when your upper body is bent forward. Its weight is shifted onto a smaller area at the front of the vertebra, and the pressure increases.

The change in curvature in the spine is also accompanied by increased tension in the ligaments. When you sit and lean forward, the front parts of two adjacent vertebrae move closer together while the rear parts move away from one another (see Figure 3). The ligaments binding the vertebrae at the rear (called the **posterior ligaments**) tighten to prevent the vertebrae from separating as they move apart. If the bending is

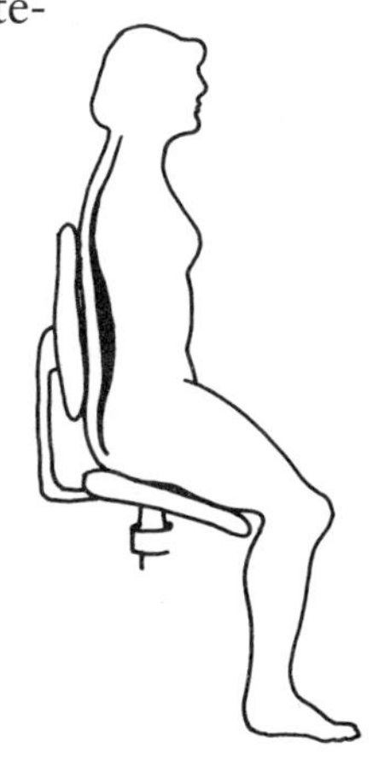

Figure 4a.
Standing erect puts a normal amount of pressure on the lumbar (lower back) discs; sitting in a chair with a declined seat pan minimizes the increase in pressure.

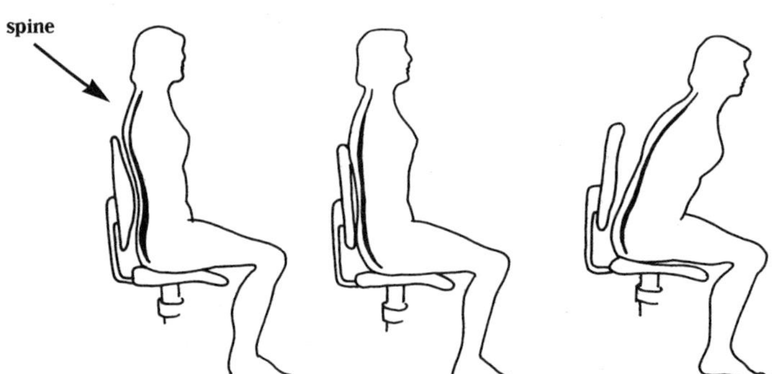

Figure 4b. The shapes of the spine and pressures in the lumbar discs while sitting in other common postures in the computer workplace.

sharp or is maintained for a long time, it gradually stretches and weakens the ligaments. This can easily happen when you sit and work at the computer with the trunk or neck bent sharply forward. Figure 4 shows the shapes of the spine in some comfortable and stressful postures.

POINTS TO REMEMBER

- Muscles end at tendons, and tendons attach to bones.
- Muscles generate forces, and tendons transmit forces across joints.
- Ligaments are strands that bind the bones forming a joint. Ligaments also prevent extreme bending of a joint.
- When a joint is bent sharply, the ligaments contract strongly to resist the bending.
- When a joint is kept bent, as when bending your neck while working at the computer, the muscles crossing it are held in a contracted state, which can easily lead to fatigue.
- Your spine consists of a chain of bones, called vertebrae, separated by compressible discs.
- Discs cushion the weight of your upper body on your spine. They absorb shocks from motion to prevent your spine from breaking.
- Your spine curves forward in the neck and lower back areas and backward in the chest and hip areas. The curves add strength to the spine and help maintain balance.

- Flattening of the curves builds up pressure in the discs of the spine.
- Bending your neck or trunk sharply when sitting flattens the spinal curve in that region of the spine. Your neck and lower back are then subjected to great pressures.
- The pressure increase will be felt as a tightness in that part of the body and eventually as aches and pains. Over a period of years, this pressure may even cause degeneration of the spinal discs.
- The sharper you bend forward while sitting, the greater the increase in pressure in the spinal discs in your neck and lower back. The tension in your muscles, tendons, and ligaments will also increase significantly.
- An increase in your upper-body weight also increases the pull tension in your neck or back muscles, tendons, and ligaments, as well as increasing the pressure in your spinal discs.

CHAPTER 8

THE BENEFITS OF SITTING AT WORK

Sitting may come as a habit, as an unconscious action performed by all human beings. It has always been regarded as something that gives the body rest and produces a general feeling of comfort. The phrase "sit down and make yourself comfortable" is common in our society. In ancient times, people sat when they wanted to rest their bodies. They probably used large rocks, fallen tree trunks, and other natural elevations as seats. They probably discovered that tired legs could be rejuvenated if the body rested for some time on its buttocks and that leaning their back on an object rested their whole body. At that time, they probably sat under trees and braced their backs against a tree trunk. Later they may have also discovered that certain work, such as weaving, could be performed better and for longer periods while sitting. However, in those days, it was unlikely that human beings sat at work for long hours. They were certainly not subjected to rigid workplace discipline, as we know it today.

In our society, sitting is not merely a way to rest the body. It has evolved into a social custom in the home and public meeting places. In schools and other institutions, sitting has evolved into a method of enforcing order and even authority over those who are supposed to sit. In the workplace, it is a method of enhancing job performance. But there are few jobs where people sit for as long a period of time without breaks as in the modern office.

The average office worker spends about seven to eight hours a day sitting at work. In a fifty-week work year, this amounts to 1,750 to 2,000 hours of sitting. Add that to the hours spent sitting at home and you get a sizable proportion of a person's lifetime. People in the developed world spend about two-thirds of their adult life sitting. Sitting is so pervasive that Swedish orthopedist A.C. Mandal, who has made comprehensive studies of its anatomical and physiological effects, has suggested that human beings in the industrialized world should be called Homo sedens (sitting man) instead of Homo sapiens (wise man).

There is a special concern about the effects of sitting at work on the body because of increasing complaints about aches and pains in muscles and joints. These complaints are not new, but they have increased markedly since the advent of desktop computers in the office. To truly understand why this is so, we must know what happens in the body when we sit. We should understand why sitting, which is necessary for work, can be harmful to our health. We must understand why people adopt certain postures and whether those postures are appropriate for their work and their health. Finally, we must learn how to sit to prevent harming our bodies.

The most common working postures are sitting and standing. Of the two, sitting has become the most common. It has become so automatic that we are hardly aware of its benefits to the body. Sitting may seem like the natural or most sensible thing to do in many situations, but this is precisely because the body derives benefits from doing it. Your brain is aware of these benefits and has developed automatic responses so that we sit without thinking about it when our body needs rest.

I have emphasized earlier that sitting and working are associated with aches and pains in the neck, shoulders, and back. Why then do we prefer to sit and work? To answer this question, we should compare sitting to standing.

- Sitting provides rest for your body, especially your legs and trunk. Since almost all workplace chairs have backrests, sitting also provides rest for your back. These are perhaps the most obvious reasons for sitting.

- Sitting provides better balance and stability for your body in general. When sitting, you are less likely to lose balance and tip over and more likely to maintain a precise posture for long periods. It is also the best posture for situations in which your arms must exert precise control over something. Using a computer keyboard is an example of this kind of controlled action. The fingers must depress small keys that occupy exact positions. The contact surface of an alphabetic or numeric key is approximately rectangular, with dimensions of only 9 x 12 millimeters. Since this is such a small area, your arms must be stable when your fingers are positioned over the keys.
- Sitting is better than standing for long periods of work, especially if you must maintain the head in a fixed position. In computer work, you have to read from the screen or documents for long periods, often intensely. During that time, you must keep your head still so your eyes can remain focused on a particular position on the screen or document.
- Sitting is suitable when it is not necessary to exert great muscular forces. Otherwise postures such as standing or stooping are better. Standing gives the body better leverage and greater ranges of motion. Since operating a computer keyboard requires relatively weak finger forces, sitting is usually best.
- Finally, sitting reduces the tendency for your legs to become swollen. In some people, especially women, veins in the legs and feet may become distended from prolonged standing due to the accumulation of fluid.

We take sitting so much for granted in our society that we are hardly ever conscious of these advantages. However, there are also disadvantages. They are related to the length of time that we sit and the postures we adopt. A job in a computer workplace forces you to sit and work, even when your body cries out for relief from aches and pains. You are not totally free to take breaks. Like other office workers, you become a prisoner of your chair, with little freedom to get up and move about. Some computer jobs are so demanding that you may not even be able to shift on your seat as frequently as your body calls for it.

Physical Stress from Sitting

If you must sit still for long periods of time, you can easily be overwhelmed by aches and pains at the end of the workday. As mentioned earlier, computer work requires you to steady your arms to operate the keyboard and your head and eyes to read from the screen or documents. You may be able to do this comfortably for several minutes but not for as long as six to eight hours every workday, as is typical in many computer jobs.

From an engineering point of view, sitting is a contest between the force of gravity and the forces generated by muscles within the body. Gravity tries to pull your body downward while muscles oppose the pull and allow you to maintain a particular posture. The strength of gravity's pull depends on the mass of your body, and the effect of the pull on the body depends on your posture. Heavier people are pulled more strongly by gravity than lighter people. And sitting with your trunk bent forward imparts greater mechanical stress and strain in the lower back and neck. The stronger the pull of gravity, the greater is the tension in the muscles, tendons, and ligaments and the greater the pressure in joints such as the vertebral discs. Over prolonged periods, the strain will accumulate and will be felt first as fatigue and later as pain.

We all know from personal experience that certain body postures produce more discomfort, aches, and pains than others. In these postures, gravity's pull produces a stronger effect on the body than in other postures. The structures that oppose gravity—muscles, tendons, and ligaments—work harder to nullify the effects. Their contractions are much stronger. Joints are also subjected to increased mechanical pressure. Such postures are called **constrained** postures. Common examples of constrained postures at the computer workstation include bending your trunk or neck forward and downward and extending your arms. Constrained postures may be tolerable for short periods, but they should be avoided for prolonged working periods. However, before you can learn to avoid them, you must understand why your body is forced into them.

At work, constrained postures are not freely chosen. If they seem natural to you, it is probably because they have become part of your daily habit. These postures are imposed on you by work conditions—the design and arrangement of the furniture and other work equipment, the method of work, and the lighting. Any attempt to correct these unhealthy sitting postures should, therefore, deal directly with the conditions that produce them.

The feeling of comfort or discomfort when sitting depends not only on the geometric configuration or degree of awkwardness of your body but also on how long a particular posture must be maintained and the freedom to shift and change it. A posture that is comfortable in the first few minutes can become uncomfortable or even painful if maintained for a long time. No posture can be maintained for several hours at work without discomfort, no matter how comfortable it may feel at the beginning.

We were always taught that sitting up straight was the proper way to sit. This was emphasized from the early days of elementary school and by our parents. Even in adulthood, we try to sit up straight when we want to look good, for example, at a job interview. In the upright posture your body assumes certain right angles, your foot is flat on the floor and horizontal, your shank (the part of your leg between the knee and ankle) is vertical and at a right angle to the foot, your thigh is horizontal and at a right angle to your shank, your trunk is vertical and at a right angle to your thigh, and your head and neck are approximately in line with your trunk and vertical. This is the **right-angled sitting posture**. Traditional chairs were designed to force our bodies into this configuration by fixing the seat and backrest at a right angle. But how good is this posture for our health?

Some seating experts, such as A.C. Mandal in the 1970s and 80s, have argued that the right-angled sitting posture has no scientific justification. It is unhealthy for the spine because it increases the pressure in the lumbar discs. So, from an orthopedic point of view, it is undesirable. Some experts have also pointed out that when left to choose our sitting posture freely, we hardly ever sit in the right-angled posture for long. Most of us tend to slouch a little after a while. How then did the right-angled sitting posture originate and how did it become so widespread?

Many experts believe that the right-angled posture was adopted from those depicted in the statues of ancient Egyptian pharaohs. Theirs was, however, not a working posture but one used for defining their authority. Another likely reason is that this posture conveys a sense of being healthy. It contrasts with a bowed upper body, which is associated with old age and physical weakness. Unfortunately, as Mandal pointed out, up to the middle of this century no one seemed to have questioned the usefulness of the right-angled posture in the workplace or the seats that have imposed it.

Whatever the reason for the use of these chairs, one thing is certain: the ancient right-angled chairs were not designed for prolonged sitting. The study of human anatomy and physiology and the study of work methods have taught us that people cannot work in the right-angled posture for long periods of time. It is certainly not the most comfortable one for the body. Since the design of the chair is one of the most important factors influencing posture, you should select your chairs carefully. Avoid chairs that force you to sit in the right-angle posture and provide little opportunity for changing position.

The Least Stressful Sitting Posture

Since posture determines the level of mechanical stress and strain in the body, it makes a great difference whether you bend your trunk forward, backward, or sideways; whether your head is erect or bent in a certain direction; or whether your arms are close to the body or extended outward. There is one posture that imparts minimum mechanical stress in the body. Knowing it helps us to understand why the right-angled posture is stressful and how to reduce stress and strain when sitting. Photographs of astronauts from early space flights revealed it.

When the space programs in the U.S. and the U.S.S.R. were being developed a few decades ago, scientists became interested in the effects of **zero gravity**, or **weightlessness**, on the human body. Since astronauts were expected to visit distant regions of space where there was no gravitational force, scientists wanted to know how their body functions would be affected. One of the effects studied was body posture.

Photographs from early space flights revealed that the human body, when not affected by gravity, assumed a posture unlike any

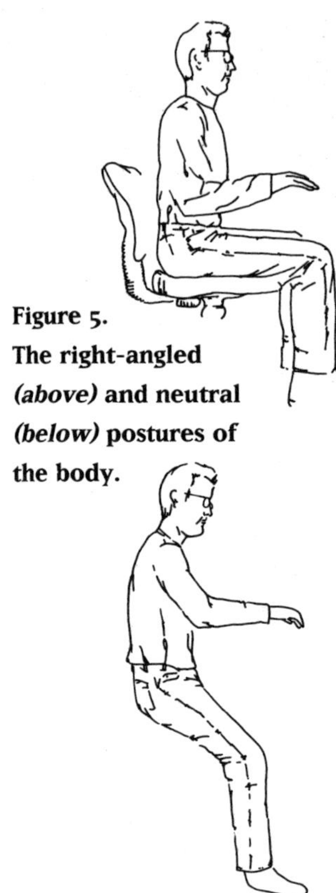

Figure 5. The right-angled *(above)* and neutral *(below)* postures of the body.

previously considered correct for sitting. Surprisingly, it bore strong resemblance to a horseback rider's posture. There were no right angles at the ankle, knee, hip, or elbow. The angles were all greater than a right angle (see Figure 5). In this posture, the structures supporting the body are in a state of balance. Since there is no external force, such as gravity, acting on the body, the muscles, tendons, and ligaments are not required to contract actively, and pressures in the joints due to muscle contraction and body mass are absent. This posture, therefore, imparts minimum stress on the body.

The minimum stress posture is also called the **zero-gravity posture** (because there is no gravitational force acting on the body), the **weightlessness posture** (because the body has no weight where there is no gravity), or the **neutral posture** (because there is no external force acting on the body).

What Mandal advocated for a minimal-stress sitting posture was similar to this zero-gravity posture. So it is no surprise that chair designers are now using this concept in the design of office chairs. In such chairs, you can adjust the seat, backrest, and arm rests to sit in the neutral posture as well as others. It is interesting to note that horseback riding has been advocated by medical experts as a means of treating certain types of back problems. The reason is, except for the wide splaying of the thighs, the body comes close to the neutral posture when sitting on a horse. If you have a chair with a tiltable seat and backrest, you may try sitting in this posture and see how it feels.

Experiencing Weightlessness

You can experience an approximation of this neutral posture if you can hold your breath long enough. The next time you are in a swimming pool, dive a few feet below water level then allow your muscles to relax. You will notice that your arms, legs, trunk, and neck immediately assume angles greater than 90° and that your body assumes the posture described above. As your body is suspended in the water, you will feel the relaxation in your muscles and joints, especially in your arms, legs, and neck. Immersion in the water cancels most of the pull of gravity on the body. This is as close as you can get to weightlessness outside a laboratory or without traveling beyond the earth's atmosphere.

Gravity and Your Body

Under the influence of gravity, the body deviates from this minimum-stress posture because its structures are subjected to a certain amount of mechanical stimulation. Gravity pulls all segments of the body downward. Muscles, tendons, and ligaments contract and pull harder as they try to oppose gravity, and joints, especially those in the spine, are then compressed. As long as the body is maintained in some posture under the influence of gravity, the pulling and compression persist. But this level of mechanical tension and compression within the body is not harmful as long as the body is not kept motionless and constrained for prolonged periods.

Gravity normally provides the stimulation for the structures of the body to grow and strengthen normally. In fact, a certain level of mechanical stimulation is necessary for the body to grow in a healthy manner. Standing, sitting, walking, running, jumping, and performing daily tasks are examples of activities that provide the correct levels of mechanical stimulation within the body. Problems arise only when the level of mechanical stress exceeds the body's tolerance for it. This happens when pull tension in muscles, tendons, and ligaments and pressure in joints are too strong, as in lifting heavy loads and sitting and working for several hours each day at the computer workstation.

Leg Muscle Contractions and Hip Motions

Sitting with your hips at a right angle flattens the curve in your lower back and increases the pressure in the discs of the lumbar spine. The reason for this is the peculiar structure of the body in the region of the hips and thighs (see Figure 6).

Your thighs are attached to your pelvis (hip bones) via the hip joints, and your hip bones are attached to your spine via joints below the lumbar spine (**sacroiliac joints**). A group of three muscles, called the **hamstrings**, runs along the backs of the thighs from the lower part of the pelvis to the back of the calves. (The hamstrings got their name because in pigs their tendons were used as strings to hang up hams.) Now this may sound strange, but your thighs cannot rotate upward more than 60° from a standing position. When you sit your thighs rotate about 60°, then your lumbar spine bends to allow the thighs to move upward the other 30° to the sitting position. You do not actually feel your spine flattening and you cannot observe the bending by looking at someone going through the motions of sitting.

The force for the flattening of the lumbar spine comes mainly from the contraction of your hamstring muscles. As you begin to sit, your hamstrings tighten and pull on the pelvis causing it to rotate backward. The pelvic rotational force then pulls on the sacrum of the spine. Since the sacrum cannot bend (it is wedged to the pelvis), the force is transferred to the next set of bones in the chain—the lumbar vertebrae—pulling on them and flattening the lordosis curve.

It is this chain of force transmission that also causes your lower back to hurt more when you stretch your legs out while sitting. This is because the hamstrings are stretched further and pull harder on the lumbar spine when your knees are fully extended. If you stretch out your legs and feel your lower back with your hands, you can detect the flattening of the lumbar spine. You will notice that the opposite action—bending your knees so that the feet are tucked under the chair—brings some relief to the lower back. This is because the contraction in the hamstrings and the pulling on the lumbar spine are reduced, and some of the forward cur-

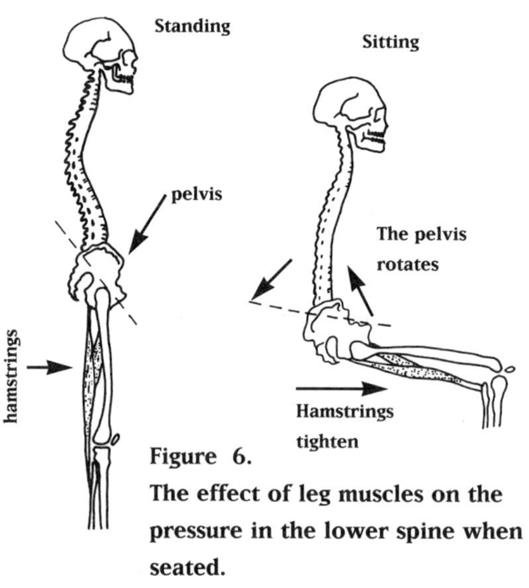

Figure 6.
The effect of leg muscles on the pressure in the lower spine when seated.

vature in the lumbar spine is regained. This is why some people occasionally tuck their feet under the chair and push their lumbar spines forward when their lower backs ache.

Preventing Pains from Pressure in the Spine

People tend to work better with their hands when they sit upright instead of reclined backward. Despite the introduction of reclinable backrests in chairs, many workers prefer to sit upright. Different methods have been proposed to sit upright and maintain the lumbar curve in the spine, but none is perfect. Some of them even create new problems. Here are the two most popular ones you can try:

- Sit on a downward inclined seat so that the thighs are lowered. This enables an open hip angle (greater than 90°) to be maintained while the trunk is in the familiar upright position. However, there may be a tendency for you to slide down off the seat. To prevent this, you could plant your feet firmly on the floor. But this increases leg muscle work and reduces your mobility. Using a seat cushion with reasonable frictional resistance helps prevent sliding but is seldom perfect.
- Sit upright with your thighs horizontal, and use a lumbar support. A lumbar support is the lower part of the backrest that has a forward curve, which fits into the curve of the lower back to maintain the lumbar lordosis. To be effective, the support must make firm contact with your lower back.

The Sitting Bones

The buttocks are remarkably adapted for bearing the weight of the body. They contain a large percentage of fat and connective tissue that cushion the pressure on them when sitting. But even this feature is not perfect. When sitting, the weight of your body is not necessarily distributed evenly over the whole buttock-seat contact. There are two projections in the lower part of your pelvis called the **sitting bones**. They act like pivots. They support the weight of your upper body, and they make closer contact with the seat surface than the rest of the pelvis does. The blood vessels and tissues under these bones are, therefore, under greater pressure than those in other parts of the buttocks. In extreme cases, like sitting on a hard surface, the blood vessels may be squeezed so much that the flow of blood in the buttocks and legs becomes restricted, leading to numbness. Your legs may "fall asleep." This is why you should not sit on a hard surface, like wood, for long periods of time. Use seats with cushions.

POINTS TO REMEMBER

- Your body derives many benefits from sitting. It gives you rest, balance, stability, and control for hand work, frees your feet for operating controls, and reduces the tendency for foot swelling.
- Your sitting posture determines the amount of strain in your muscles and joints. It may be responsible for aches and pains in your neck, shoulders, lower back, and legs.
- A posture that is comfortable for the first few minutes of sitting may become intolerable after a few hours.
- The least stressful posture, the one that imposes minimum overall stress in your muscles and joints, closely resembles the horseback rider's posture without the wide splaying of the legs. Some seating experts recommend sitting in this posture, with the necessary variations to permit working.
- Some ergonomically designed chairs can be adjusted to achieve this posture.
- Avoid postures that flatten or reverse the forward curve in your lower (lumbar) spine.
- The traditional upright sitting posture imposes great stresses on the lower back by keeping the hips at right angles and flattening the curve in the lower back.
- Seating experts recommend opening out the hip angle or using a lumbar pad to maintain the curve in the lower (lumbar) spine when sitting.
- Stretching the legs out for significant periods of time builds up pressure in the lower back. Leg muscles pull on the pelvis, and the pelvis pulls on the lower spine.
- The best sitting posture for relaxing is not necessarily the best for working, and vice versa.

CHAPTER 9

COMPUTER FURNITURE

You have learned that the way you sit in front of a computer determines whether your body will ache, what part(s) will ache, and how severe it will be. A little thought will make it clear that your sitting postures are determined by the design and arrangement of computer furniture, the characteristics of your computer hardware, the nature of your work, and your physical characteristics. Any attempt at influencing posture by modifying one of these work factors should also take into account the others. Design features include shape, size, angular configuration, and material used. The view of ergonomics is that the chair, the desk, the computer hardware, and the person form a closely integrated system. None should be treated in isolation.

Matching the Chair and Desk

Your chair and desk are, without doubt, the two most important pieces of furniture at your workstation. The chair is more important, because it is in direct physical contact with the body and because greater variation in posture can be achieved by adjusting its features than by adjusting those of the desk. So most of the discussion on furniture will be centered on chairs. The desk will be mentioned wherever it completes the picture of the computer workstation or helps explain your body's reactions.

There is no such thing as a perfect chair. Comfort has a time limit. A chair that may be comfortable for ten minutes may become intolerable after several hours. Also, the comfort of a chair depends on the design and arrangement of the desk in front of it. No chair can prevent aches if it is used with a poorly designed desk for long periods of time. And a chair that is suitable for one particular desk may be unsuitable for others.

The chair and the desk, therefore, cannot be considered independently when designing a workplace. Any change in one may require a change in the other. For example, if you are a short person, you may find that a particular seat is so high that your feet cannot rest comfortably on the floor while sitting but that the desk height is acceptable. You may then lower your seat height to get your feet comfortably on the floor. But alas! You now discover that you must raise or extend your arms uncomfortably to use the keyboard because the desk has become too high.

Work Furniture as Tools

Chairs and desks in your office are tools for your use the same way a screwdriver or a wrench is a tool to a maintenance technician. These tools help you perform your work in a simpler and more effective way, and how well they are designed will determine how effectively you work.

The design, fit, and comfort qualities of these tools influence how long you can work before experiencing muscular fatigue. If you regard your chair as a tool, then its importance is automatically in the forefront of your mind. You should, therefore, try more earnestly to understand how it works and treat it just like any other tool. Learn how to adjust and set its features, such as seat height, backrest angle, backrest height, etc., so that you can sit and work comfortably. Employers should also provide for maintenance and repair of chairs when necessary, just as they do for other tools. A flawed chair can be just as harmful to your health as a flawed hand tool.

Within the computer workplace, the chair may even be regarded as important as the computer itself. If it is poorly designed, it will most likely make you miserable at work. It will also adversely affect the quality of your work and your productivity.

Furniture, Productivity, and Health

One of the main reasons the chair and desk have not been regarded as important office tools is that their effects on productivity and health are difficult to evaluate. There is still no accurate and reliable method for measuring the effect of a change in furniture on workers' health or productivity. That kind of change cannot be observed overnight. It is a long-term effect, and there are too many other variables that confound it.

It is also possible that the scant respect traditionally given to workplace furniture is due to the incorrect view that it merely gives workers rest. Actually, office furniture can determine the quality of the end product. So little respect has been given to workplace chairs that all kinds of objects have been substituted for them. Even today we see employees in some workplaces sitting on boxes or crates, instead of chairs or other suitable kinds of seats, when they are not working.

In the workplace, we are expected to understand how our computers work, how to use various types of computer software, and how to evaluate the quality of our own work. But what about health? Have we been encouraged to learn how our health can be affected by our work furniture? Hardly. But until we understand the relationships between health and computer-workplace conditions and between productivity and health, we will never be able to appreciate the full importance of the physical aspects of our workplace.

Cost of Operation

It is difficult to measure exactly how many dollars will be gained in productivity or saved in health costs for every dollar spent on new furniture. Managers and supervisors are often hard pressed to convince their bosses that significant gains will be made by the introduction of new (and often costly) furniture. But discomfort, aches, and pains in the workplace reduce your productivity and quality of work. Under this kind of physical strain, you tend to make more errors or work more slowly to maintain accuracy. When the brain is busy dealing with strain, people are often unable to concentrate. In addition, you may be forced to take more pauses to rub your aching muscles. But many employers still don't realize this, and many people still don't see the link between the design of work furniture and productivity. Many of those who see the link often doubt that the problems can be solved cost effectively. This is where education in workstation ergonomics comes in.

The functions of computer-workplace furniture should be analyzed carefully. You need to know how furniture can reduce or eliminate your aches and pains, and management needs to know how the money spent on improved furniture fits into the benefit-cost equation of the company's operations. Benefits include improvement in efficiency and productivity, as well as reductions in costs for sick leave, medical treatment, and rehabilitation.

It is not difficult to estimate the worth of an ergonomic chair in the workplace. Professor Stephan Konz of Kansas State University, a leading ergonomist in the U.S., has provided a simple numerical example of such an assessment in his college textbook, *Work Design: Industrial Ergonomics*. The following example is a slight modification of that model.

Assume that an ergonomically designed chair costs six hundred dollars and that it would last for ten years, after being used for forty hours per week (or two thousand hours per year). In ten years, the chair would be used for twenty thousand hours, and the cost of using it would be three cents per hour. Assume also that the alternative is a chair without good ergonomic features that costs about one hundred dollars. Again, if we assume a "chair-life" of ten years, the cost per hour for using it would be one-half cent. In other words, the ergonomically designed chair would cost about two-and-a-half cents per hour more. This is equivalent to increasing the wages of an operator by two-and-a-half cents per hour. This is incredibly small compared to a typical wage of about ten dollars per hour for a skilled computer operator. It is a mere one quarter of one percent of a worker's wages. The increase in cost of operation to the company is negligible, making it an excellent investment. The comfort provided by an ergonomic chair delays the onset of fatigue and can reduce the need for long or frequent breaks for relieving muscular fatigue. The gain in productivity can be enough to recover the extra cost of the better chair. Well-designed chairs with adequate ergonomic features cost anywhere from five hundred to one thousand dollars per chair, but companies are often able to get a discount of a few hundred dollars per chair if a sizable quantity is ordered.

Selecting Furniture

Most people know enough to understand that the prudent choice of a chair or desk can make the difference between an unbearable backache and mild discomfort. This is the kind of knowledge you gain from hard experience. Yet many people find themselves using furniture at work that never seems comfortable. You might wonder how certain furniture ever made its way into the computer workplace. Who chooses the furniture? What considerations are applied in making the choice? These are important questions, and anyone who works at a computer should learn the answers.

Office furniture, especially the chair, has evolved into a complex machine. In the modern office, the wise selection of furniture has become almost a scientific skill, and it should not be left to someone who lacks this skill. The one who selects furniture should have at least a basic knowledge of the ergonomic aspects of seating and workplace design. Unfortunately, in many workplaces the task of selecting new furniture is given to people who manage money rather than the people who understand the relationships between how people sit and how they work. Selection is often based on features such as cost and physical appearance (matching color, shape, and other aesthetic variables) instead of comfort, health, safety, and productivity.

We cannot select furniture in the same way that children select their breakfast cereals. We should not allow our powers of reasoning to be destroyed by the impact of aesthetically appealing, brightly colored pictures of attractive furniture in magazines. Instead, we should focus more on the importance of health and comfort than how attractive the workplace will look. Work furniture is for the benefit of workers, not for the sensory gratification of others. The folly of making decisions with such flawed reasoning is likely to be discovered only after the investment of thousands of dollars in furniture, when replacement may be considered too costly. Regardless of what other considerations are deemed appropriate, the choice of workplace furniture should always be based on enhancing workers' health and productivity. When this is achieved, both you and your employer will be happy.

What Is Ergonomically Designed Furniture?

One of the most popular terms used in the advertisement of office furniture is **ergonomically designed**. It is a catchphrase that furniture manufacturers and distributors use for boosting sales. But this claim must be examined carefully because it is not always accurate.

In a broad sense, ergonomically designed work furniture is furniture that conforms to the physical characteristics (**anthropometry**) of the human body as well as the characteristics of the job. Such furniture should allow you to sit comfortably and perform tasks efficiently. Ergonomically designed also means having adjustable features to fit people of a wide range of body sizes and characteristics and to allow any person to vary his posture. Any given workplace has people with a wide range of heights and sizes. When using an ergonomically designed chair, a short person should be able to lower the seat height and a tall person should be able to raise it, and any user should be able to change features such as the backrest or seat inclination to change posture.

In chapter 11, you will learn how to evaluate the ergonomic features of your furniture. If you are contemplating buying new furniture or making changes to the arrangement of your present ones, you should read that chapter carefully.

POINTS TO REMEMBER

- Your chair in the computer workplace is a tool in the same way that wrenches, pliers, and screwdrivers are tools to the mechanic.
- For work, select a chair that suits your particular job and makes you comfortable. Do not be carried away by features such as color and shape.
- The construction of your chair should match your desk to enhance your work efficiency and maintain your health.
- Learn about the various adjustable features in office chairs before you buy one. (See chapter 11.)
- An ergonomically designed chair is one that can be adjusted to (1) accommodate people of a wide range of body sizes, (2) allow you to work productively and efficiently, and (3) allow you to vary your posture.
- A chair that costs six hundred dollars is equivalent to a salary increase of only two-and-a-half cents per hour per worker if you use the chair for ten years.

CHAPTER 10

BODY SIZE AND COMPUTER FURNITURE

There are many things to consider when choosing an office chair. Comfort, orthopedic qualities, usefulness or relevance, cost, aesthetics, durability, shape, availability, and hygiene as well as how well it matches other furniture, the walls, and the carpet are all important factors. Selecting a chair based on only one important quality (e.g., low cost) can lead to problems for the person who uses it. Each quality satisfies a need, but some needs are more important than others and some qualities mean different things to different people.

It is rare that employees are allowed to select their own chairs, but you can seek help from your supervisors, Facilities Department or Employee Health Department and voice your concerns about any work furniture that has been assigned to you. Large companies now tend to have management personnel who are knowledgeable in furniture and workplace design, and your concerns are likely to be given serious attention. From what you learn in this book, especially in chapter 11, you will be able to explain your problems and needs concerning work furniture to the professionals in this department with ease.

Different people may have different priorities when selecting a chair. Someone who wants to avoid aches and pains at the end of the workday may consider orthopedic and comfort qualities the most important ones. A visitor to the office may consider color or shape most important. Employers, on the other hand, tend to consider cost the most important factor. However, attitudes toward office furniture and workstation design are gradually changing. Many employers are now placing emphasis on comfort and orthopedic qualities of chairs. This is due to a better recognition of the effects of furniture on workers' health and productivity and to a determination to minimize the great costs of health problems caused by working at a poorly designed workplace.

Employers should not worry about how much they should spend on ergonomic furniture for computer workplaces because the expenditure can be justified on a benefit-cost basis. Buying chairs is a cost, but the alternatives—lowered productivity and efficiency, medical treatment for cumulative-strain illnesses, workers' compensation, absence from work, and training of new workers to replace disabled ones—incur greater costs. Scientific evidence indicates that ergonomically designed furniture in the computer workplace can significantly reduce costs related to musculoskeletal problems, and the money saved can easily offset the cost of furniture.

Measuring the Importance of Furniture Features

The more qualities that are considered in the selection of furniture, the more difficult the choice becomes. The important question is, how much importance should be given to the various qualities? Can anyone say, for example, whether utility is twice as important as durability? Or three times as important? These are difficult concepts. Except for cost, the importance of these factors cannot be measured accurately.

Seating experts do not yet have an objective method for accurately determining the relative importance of each quality of a chair. However, there are certain general guidelines that can be followed. The most important one is that the person who will use a chair at work should use it over a trial period before making a decision to accept it. How long should this trial period be? Again, there is no definite period but several days will certainly provide enough feedback. The chair should be used under all conditions in which it is expected to be used in the workplace. If it would be shared by different workers (in different shifts, for instance), then it should be tried out by each one—especially if they are different sizes. No one should believe that sitting for a few minutes in a chair is enough to judge its appropriateness for the workplace. And remember, the advice of a salesperson should be examined carefully because they don't always understand the importance of ergonomics. Someone who is about to spend tens of thousands of dollars on office furniture and is not knowledgeable enough to evaluate it should seek the help of a qualified ergonomist. A qualified ergonomist is knowledgeable in the nature of work, the structure and functions of the human body, how the body responds to physical and mental stress from work, how the body and mind react to stress, and how work can be designed to minimize stress.

Body Size and Sitting

The furniture in the office must fit your body size and be suitable for your job. Otherwise, certain functions of your body may not work at their best. To prevent aches and pains, the most important of these functions is the ability of the muscles and joints to generate and withstand tolerable levels of forces and, simultaneously, maintain body balance and stability in a particular posture.

Below is a list of some important body-size dimensions that determine how well you fit your furniture. The values of the measurements are not given here, but you can find more details in the next chapter.

- Knee height: the height from the sole of your foot (or floor) to the top of your knee. This determines the height from the underside of your desk to the floor.
- Popliteal height: the height from the sole of your foot to the underside of your knee. This determines the height of your seat.
- Buttock-popliteal length: the length from the rear of the buttock to the underside of the knee. This determines the depth (length) of the seat of your chair.

- The thickness of your thigh. This determines the distance between the seat and underside of your desk.
- The width of your hips when sitting. This determines the width of the seat and influences the separation distance of the two armrests.

Scientific evidence indicates that many problems of comfort, health, and work productivity are due to a mismatch between your body size and the size, or design, of your work furniture, especially your chair. When work furniture is either too large or too small for you, your work posture becomes constrained. Your body is subjected to excess mechanical stress; and muscles, tendons, ligaments, and joints can become strained.

Severe mismatches between people and furniture often occur in workplaces where only one size furniture is available for people of widely different sizes. Management can avoid this dilemma by either providing an adjustable chair, at the minimum, or a combination of an adjustable chair and an adjustable desk. Assigning workers of comparable sizes to the same workstations is another useful alternative.

The range of human body size is vast. The variation is due partly to genetic control of our growth and partly to the environment and cultural customs—nutrition, exercise, type of work, and state of health are only a few of the factors involved.

According to the *Guinness Book of World Records*, the tallest person who ever lived was 8-feet-11 and the shortest was 2 feet. This is a wide range—6-feet-11 to be exact. Although it is unlikely you will find people spanning such a wide range of heights in your workplace, it is not uncommon to find workers spanning a 1½ foot range (e.g., 5 feet to 6-feet-6). This range is great enough to make furniture of a fixed size inappropriate. The design of furniture, especially the chair, becomes crucial. If a fixed-size chair is used, then some people will be accommodated while others will not. If only average-size chairs are available, then very tall and very short workers are likely to experience aches and pains after sitting for long periods. Providing one size of chairs was the practice many years ago when designers and manufacturers produced chairs to fit people of average height. Such chairs were tolerable at that time because most workers did not have to maintain stresses for prolonged periods. But they are not suitable for today's computerized office jobs. The nature of computer work freezes your body into one posture for many hours, and the effects of even small mismatches become accentuated. You need a chair that fits your size and allows you to change posture from time to time. You need an adjustable chair.

Adjustable chairs are widely available today. The angles and heights of the seat and backrest, as well as many other features, can be changed easily and quickly. Adjusting these features is like creating different chairs out of one. These adjustable features will be discussed in detail in the next chapter.

Designing for Adjustability

If the heights and sizes of all workers were the same, only one size chair would be necessary. However, since the range of variation of the population is wide, more than one size chair is necessary. The fact that ergonomists have recommended adjustable chairs for computer work and other seated workstations is a clear indication that the range of variation in human body size is considered too great for fixed designs.

A single unit of adjustable furniture, such as a chair, can be made to fit a large percentage of the population with simple adjustments. For example, the height of the seat pan in a modern office chair can easily be raised to suit a tall person or lowered to suit a short person by pressing on a lever below the seat pan. The height of the backrest can also be raised or lowered by pressing a lever or turning a knob.

Adjustability allows more than the accommodation of a large percentage of the population. It also accommodates a wide range of postures for a single person. For example, you can sit erect or reclined by simply changing the angle of the backrest. The seat pan angle can also be changed to produce a different angle at the hip when sitting. An office chair cannot be considered ergonomic and, hence, suitable for the modern office if it is not adjustable. Even desks are now made adjustable so that you can raise or lower their work surfaces or tilt them. And some desks now have a keyboard rest with adjustable height and tilt.

One of the most well-known examples of an adjustable-chair design outside the office is the driver's seat in a car. It can slide forward or backward and set in many different positions within that range, so that both short and tall people can reach and operate the steering wheel, foot pedals, and dashboard controls. Backrest angles in all cars and seat inclination in some types can also be changed easily to produce slight changes in sitting posture.

Nonadjustable Chairs of the Past

Before the advent of the computer workplace, nonadjustable chairs did not seem to create significant health problems, even though they came in only one or a few sizes. This was because of the nature of the jobs. Sitting jobs in the office were not as physically stressful as they are today.

Also, there were fewer and looser laws and regulations protecting people's health in the workplace. This discouraged complaints of work-related stress and strain by workers. Before the Occupational Safety and Health Administration (OSHA) was created in 1970, there were fewer avenues for workers to complain about poor workplace conditions. Many people will also point out that because our society was so male dominated there was less concern among safety professionals and lawmakers, who were predominantly males, for the health and general welfare of the office workforce, who were predominantly female.

Fitting the Whole Population

Designing for adjustability does not mean that 100% of the population will be accommodated. In fact, ergonomists and designers have learned that it is not economically feasible to design most things that depend on body size, such as a chair, to fit 100% of the population. It is more expensive to design for extremely small or large people. Take the familiar example of a car. If, for example, the range of motion of the driver's seat had to accommodate people 4 feet to 7 feet tall, think about how much more size and cost would be added to the vehicle. In the same way, the range of adjustability of the seat height of an office chair is limited. Cost and size are powerful factors in determining the limits. Designers usually try to cater to as much of the population as possible under these constraints. Throughout the years they have found that they can design for as much as 90% to 95% of the population economically.

Some chair manufacturers are now producing special chairs for very short, very tall, and very large people. They are, of course, much more expensive than the normal ones that are designed for the other 90% of the population.

POINTS TO REMEMBER

- Your chair and desk should match your body size and the nature of your work.
- People of widely different sizes, for example 5 feet to 6-feet-6, cannot all be comfortable in the same fixed-size chair. But an adjustable chair can accommodate a wide range of body sizes.
- An adjustable chair allows change of posture.
- Designers find it economically unfeasible to design work furniture to fit everyone in the population. They aim for the middle 90% or 95% of the population because designing for the other 10% or 5% is often expensive.
- The best furniture designers base their designs on statistical distributions of body segment sizes and ranges of motion.

CHAPTER 11

EVALUATING COMPUTER FURNITURE

You have learned that the way you sit may be responsible for aches and pains, especially in your neck and lower back. Your sitting posture determines the intensity of strain in your muscles, tendons, and ligaments and the pressure in your joints. You have also learned that your posture is influenced by the nature of your work, the design of your work furniture, workplace arrangement or design, and workplace lighting. A lot of attention has been given to furniture design recently. Today hardly any employer would seek to improve work conditions without considering how the furniture affects workers' health and productivity. This chapter describes how you can evaluate the different features of your computer furniture.

It would be an incomplete discussion to describe the different features of furniture design without also examining how the furniture and work equipment (especially your computer) are arranged. Comfort in a chair is not necessarily an inherent quality. It depends on the way the chair is used. A comfortable chair in one work situation may be uncomfortable in a different work situation. The arrangement of the chair, desk, computer monitor, keyboard, and document holder, therefore, affects aches and pains in the body.

You have already learned that, if a chair or desk must be used by people spanning a wide range of body sizes, it should be adjustable enough to accommodate a wide range of body sizes. If you are the only user of a chair, adjustability for body size is not critical. However, adjustability that helps you to change your posture is always necessary. Adjustability for body size depends on the size of the chair, whereas adjustability for posture depends on the angles in different parts of the chair. For example, experts recommend that the seat height of a chair for a workplace should be adjustable from 16 to 20½ inches because this corresponds to the range of comfortable sitting heights for the very small to the very tall. But if your popliteal height is 18 inches and if the chair would be used by you alone (as in your home office) then such a wide range of adjustability is probably unnecessary. A chair with a fixed seat height at 18 inches or a little less would be adequate for you. However, since you need to be able to change posture to relieve backaches, your chair backrest should tilt backward and forward.

The Human Factors and Ergonomics Society of America (HFES) has made a study of the computer workplace and has recommended, among other things, design features that computer-workplace furniture should incorporate. These recommendations, sanctioned by ANSI (the American National Standards Institute), constitute the available standards in the U.S. for computer-workplace furniture. They are referred to as the **ANSI/HFS standards** and are published in a document called "American National Standard for Human Factors Engineering of Visual Display Terminal Workstations." The smallest 5% of women and the largest 5% of men have been excluded, for reasons discussed earlier. If you need details on the standards for computer equipment, furniture, and work environment, you should consult the ANSI/HFS document.

Having a set of standards for a particular equipment, work, or work environment does not necessarily mean that all problems can be solved by strict adherence to those standards, especially when dealing with relatively new equipment or new work methods. Therefore, employers should not feel that their obligation in maintaining a healthful workplace is finished after furnishing workers with the latest computer equipment and furniture that meet ANSI/HFS standards. They must continue to monitor workers' health. The ANSI/HFS standards carry the respect of two prestigious scientific bodies. Therefore, they influence furniture design considerably and are a selling point in the furniture industry. Employers are attracted to manufacturers' advertisements stating that their office furniture conforms to ANSI/HFS standards. But you should remember that using such furniture does not necessarily guarantee freedom from musculoskeletal strain. To achieve freedom from strain, you must adjust this furniture to suit your size, work characteristics, and personal behavior.

There are other ergonomic standards in Germany, Britain, Sweden, and a few other countries. They are quite different from the American ANSI/HFS standards in some respects, even though differences in the body size in the populations of many European countries and the U.S. are not significant. These differences in standards reflect differences in opinions and methods at arriving at the standards. In other words, the international scientific community is not in agreement on some aspects of the human body's reaction to computer work. So it doesn't seem prudent to have rigid standards for those aspects of the computer work environment that we are unsure about. Do not rely solely on books or other documents that quote figures on standards. Strive to understand the science behind the recommendations. This is where the information in this book will help you.

The Chair

The chair determines your body posture more than any other equipment in your workplace. The earliest chairs could have been as simple as a single piece of wood carved out of a tree trunk. When craftsmen became more skilled, they made the backrest, seat pan, and legs as separate pieces, then fitted them together. Naturally, this allowed greater variation in the shape and size of a chair. But once the pieces were fitted together, the dimensions and angles of that particular chair were fixed and your posture was predetermined. Eventually designers learned to design the backrest so that it could be tilted as well as moved up and down.

Similar improvements were made in the seat and other parts of the chair, and new features such as armrests and inflatable lumbar supports were added, leading ultimately to the present ergonomic chair designs.

There are so many adjustable features in some types of ergonomic chairs that you can become scared of them. But some of these features may be trivial and make no real contribution to health or comfort during work. In this chapter we will discuss only those features that are of practical significance—the ones you must look out for when you are selecting a chair for your workplace (see Figure 7).

THE SEAT PAN

The seat pan of a chair is the part you sit on. It forms the main contact with the body, and its functions are to support the weight of the body and maintain the body in a stable and comfortable position. It is commonly referred to as the seat. The features of the seat that determine support, stability, and comfort are width, depth, shape, angle of inclination, contour, firmness, and the cushion's resistance to sliding. These seat features must be examined scientifically in order to determine the suitability of a chair for a particular person or group of people.

SEAT SIZE

The dimensions of the seat determine the pressure on your buttocks and thighs. They influence how long you can sit comfortably. Have you ever sat on a very small surface, such as a strip of wood about one inch wide, or on the edge of a bench? If you have, you probably remember the great pressure on your buttocks and the ensuing discomfort and pain. Sitting on a larger area reduces the discomfort considerably because the pressure is reduced.

The area of contact between the seat and the body is directly related to the pressure on the buttocks and, hence, to the level of discomfort. When you sit, your buttocks and upper thighs bear the weight of your body above the legs. On average, the weight supported by the buttocks is about 65% of your body weight. (Your legs account for 32% of your total body weight.) So, if you weigh 200 pounds, your buttocks will be supporting about 136 pounds! The exact proportion supported will, of course, depend on the distribution of your body weight and on your posture. If you have a relatively large upper body and small legs, the proportion will be greater than 65%, and if you have a small upper body and large legs, the proportion will be less than 65%. If you lean back on the backrest or lean forward, less weight will be supported by your buttocks than if you sit upright.

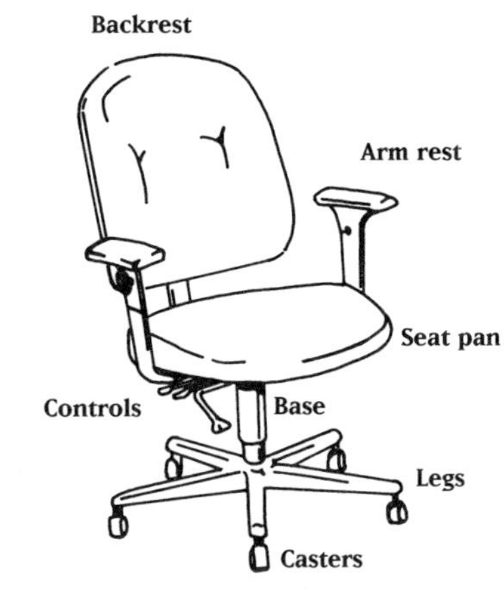

Figure 7.
A modern ergonomically designed chair, showing the most important features.

The weight supported by your seat is spread around the area of the seat in contact with the body. With greater contact area, your upper-body weight would be spread around more widely and you would feel less strain. In other words, it is not the total weight that matters but the weight per unit area of seat in contact with the buttocks. If you have ever studied physics in school, you would recognize this concept as **pressure**. In some parts of this book, it is referred to as **mechanical pressure** in order to distinguish it from psychological pressure.

The minimum pressure is achieved when the buttocks make maximum contact with the seat, and this occurs when the seat is sufficiently large. Let's call the area at which maximum contact is made the **optimal area** and the corresponding pressure the **optimal pressure**. A smaller-than-optimal area will produce greater-than-optimal pressure, but a larger area will still produce optimal pressure. A seat pan that is too small for the buttocks, or less than the optimal size, will place greater pressure on the buttocks than is necessary. The smaller it is, the greater the pressure will be. This problem occurs with people with very large buttocks (for example, obese people) who use seats designed for much smaller people. They may be too embarrassed to request a chair with a larger seat, but management can avoid this problem by providing chairs with the largest size seats available in the market for these employees. Some chair manufacturers now sell chairs with large seats designed for such people.

THE EFFECTS OF A NARROW SEAT

A seat that is too narrow may cause increased pressure on the buttocks, restriction of movement, and instability of the body. A narrow seat restricts the area of the buttocks over which body weight is distributed. Part of the buttocks may even overlap at the sides of the seat. The pressure on the buttocks will be greater than optimal, but, if the overlap is slight, this extra pressure may not cause a problem. For a large overlap, especially if the seat has no cushion, the pressure increase may be enough to compress blood vessels and other tissues in the buttocks and restrict the flow of blood. Numbness, cramps, or pain may then be the result from prolonged sitting.

Greater problems may be caused from the restriction of movement that the small seat pan imposes on your body. A narrow seat will impede getting into and out of the chair if the chair has armrests. Fixed armrests tend to be as wide as the seat pan. Therefore, large (especially obese) people can have difficulty fitting between them and onto the seat of a narrow chair. Fortunately, armrests that can be adjusted to accommodate broad people are becoming common in ergonomic chairs.

When your body is not stable on a seat, your legs are called upon to make firmer contact with the floor to compensate for the deficit and maintain balance. This can produce early leg-muscle fatigue. Available office chairs are typically broad enough to accommodate most people. Only people with very large buttocks may find it difficult to secure chairs with broad enough seats.

A DEEP SEAT

The distance from the rear edge to the front edge of a seat pan is called its **depth**, or length. A seat should be neither too deep nor too shallow for you. It must be deep enough to allow contact with your buttocks and upper thighs. But, if it is too deep, its front edge will press on the backs of your knees and thighs. If it is too shallow, it may increase the strain in your lower back, buttocks, and legs and may cause instability. Fortunately, shallow seat pans are rare among office chairs, but seats that are too deep to achieve comfort are common. A seat that is too shallow will not permit enough contact with your buttocks and legs, which increases the pressure on your sitting bones and buttocks tissues.

A seat that is too deep restricts the circulation of blood in your legs and produces numbness or cramps because its front edge presses onto the backs of your knees or thighs. Blood vessels will be compressed and blood circulation inhibited. To avoid mechanical pressure, you may be tempted to shift forward and sit on the edge of your seat so the backs of your knees and thighs clear the front edge of the seat.

Sitting on the edge may solve problems at the backs of your thighs and knees, but it can create new problems in other parts of your body—the lower back, buttocks, and legs. The pressures in your lumbar spine and buttocks may increase, and your leg muscles may be forced to work harder. When you sit on the edge of your seat, your back makes no contact with the backrest and your whole upper-body weight rests on your spine, making pressures in the lumbar spine and on the sitting bones greater than when you lean on the backrest. Because sitting on the edge provides a smaller sitting surface, the pressure on the buttocks also increases. In normal sitting, contact between the seat and upper thighs enhances body stability. This extra contact is lost when you sit on the edge of your seat. You sit mainly on your sitting bones, and your body becomes less stable. Your back and leg muscles are then forced to work harder to compensate for this loss. You must also plant your feet more firmly on the floor to enhance stability. Some people tend to push their chairs backward and lean forward when they sit on the edge of their seats. In this posture, the back muscles are forced to work harder to maintain the bent trunk, and compression in the spine increases.

Another disadvantage of a seat that is too deep is that it does not allow you to tuck your feet under your chair easily. Many people tuck their feet under their chairs for short periods to relieve pressure in the lower back. But there must be enough clearance between the back of the knees and the front edge of the seat to do this. Ergonomists agree that this clearance should be about two to four inches. But you can determine if your seat provides enough space for you without performing any measurement. Sit with your back firmly on the chair's upright backrest. Now, tuck your feet under the seat. If you feel the front edge pressing onto the back of your knees, the seat is too deep for you.

WATERFALL SEATS

You will notice that the seat pan of a modern office chair has a downward curvature at its front edge. This is a special design feature intended to protect the tissues in the backs of your knees and thighs against excessive compression. This kind of seat pan is called a waterfall seat pan. By minimizing tissue compression, the front curve helps maintain the free flow of blood in the underside of the legs and, as a result, prevents numbness, cramps, and pain.

ANSI/HFS standards state that the seat depth should be between fifteen and seventeen inches for the American population. Only extremely short people may find these dimensions uncomfortable.

SEAT CUSHIONS

The balancing and stabilization of your upper body when seated is the job of the muscles in your trunk and legs. But the amount of work performed by these muscles depends on how your body is positioned on the seat by the sitting bones (see Figure 8).

Your sitting bones act as pivots for your body, and, because they are small, they are subjected to great pressures. The exact amount of pressure depends on how the body weight is distributed over the seat. On a hard seat surface such as wood, your body weight is not well distributed over the seat and there is little cushioning effect so the pressure on the sitting bones is great. The tissues and blood vessels in the buttocks are compressed, and blood flow in the buttocks and legs can be impeded. During prolonged sitting, this can easily create numbness in the buttocks and the legs.

The fat, muscles, and connective tissues in the buttocks have a cushioning effect. The more of these tissues you have, the better protected you are from the pressure on the sitting bones.

The widespread use of cushions on the surfaces of seat pans is the furniture designer's way of reducing this pressure. In addition to providing softer contact with the buttocks, a cushion depresses under the weight of your body and assumes the contours of the buttocks. It therefore increases the area of contact with the seat pan and distributes the upper-body weight over a larger area on the buttocks. Consequently, the pressures on the sitting bones and surrounding tissues are reduced, and excessive compression on blood vessels is prevented.

Pay attention to your clothing. Some types of undergarments can slide against your outer clothing. This will force the muscles in your lower trunk, buttocks, and legs to work harder to prevent you from sliding forward on your seat, resulting in earlier muscular fatigue in those areas of your body.

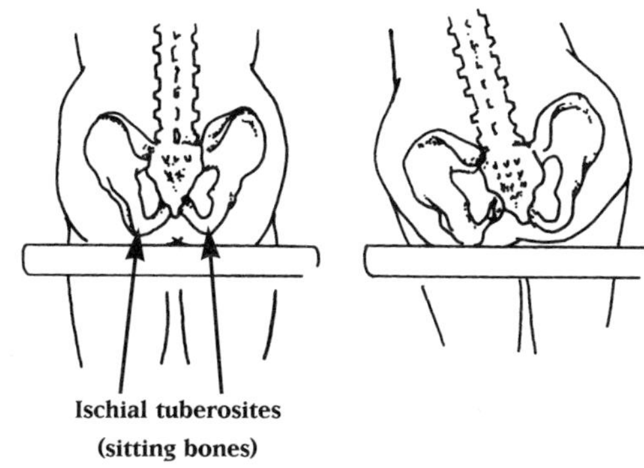

Figure 8.
Pressure on the sitting bones, blood vessels, nerves, and other tissues increases when you sit to one side, producing cramps and numbness

No seat is perfect for prolonged sitting. No matter how well designed a chair is, certain muscles, especially those in the lower back and around the hips, will eventually become fatigued. The natural response is to change sitting posture slightly so that other muscles take over the job of stabilizing the body. To achieve this, you should be able to shift sideways, forward, and backward on the seat. Different people have different reactions to discomfort, but the general method is to lift the buttocks or slide them along the seat pan. However, if the buttocks are sunk too deeply into a soft cushion, this kind of shifting becomes difficult and you may be more inclined to remain in one posture. The muscles in your trunk and legs that maintain balance can then be overworked. In other words, a seat that is very soft and initially comfortable can become uncomfortable and, perhaps, stressful when used beyond a certain time. Some degree of firmness in the cushion is necessary. Fortunately, modern office chairs seem to satisfy this requirement. Most experts on seating agree that the seat cushion should be depressed about one inch when sat on. This would make it soft enough to prevent pain and numbness in the buttocks but firm enough to facilitate shifting and change of posture.

Muscular discomfort, numbness, cramps, and pain in any part of the buttocks or upper thighs may cause you to shift on the seat and lean sharply to one side or cross your legs. But you may not always succeed in reducing the pressure in this way. Leaning sharply to one side transfers the pressure from one part of the buttocks to another. While the fatigued part gets relief the other part will be under greater mechanical stress than before, and the shifting and leaning will soon resume toward the other side.

Sitting on only part of your buttocks has another adverse effect on your body. It upsets your balance, and the muscles in your lower back, buttocks, and legs must work harder to reestablish it. Muscular work in the trunk also becomes unequally distributed when you lean toward one side. If you lean to the left, the muscles on the right side of the trunk work harder than those on the left and vice versa. Concentration of work in one set of muscles can be a cause of early fatigue in that part of the body.

The frictional resistance of the seat cover is important for stability. A surface that is too smooth, like polished leather or smooth wood, has little resistance to slipping. It is not easy to lean back on such a seat without sliding straight out and down onto the floor unless you use your feet to push and oppose the sliding. The expensive, attractive leather chairs that business executives are fond of are not the kind that you should ever use at a computer workstation. These chairs usually have highly polished leather on both the backrest and seat, which caues the buttocks to slide forward when you lean back or slouch. To prevent sliding, you would then be forced to plant your feet more firmly on the floor, increasing the work of the muscles in the trunk and legs.

Several different designs in a seat have been tried for combating forward sliding. The most popular and effective are listed below:

- A seat cushion with high frictional resistance.
- A contoured depression in the seat surface that the buttocks can fit into. However, this should not be too deep.
- A seat pan inclined slightly upward, that is, the front end tilted upward and the rear end titled downward. At the same time, the backrest can be tilted backward to get a comfortable hip angle. A fixed seat pan is usually tilted upward at five degrees, even if it has a cushion with appreciable resistance.
- A combination of the above three models.

A deep contour on the seat pan may be comfortable for a short while, but it is undesirable for prolonged sitting. It fixes you too snugly into the seat and restricts mobility. It then becomes more difficult to shift on the seat and make slight changes in posture to relieve fatigued back and leg muscles. As ergonomists say, postural fixity is not beneficial to your body.

THE BACKREST

The purpose of a backrest is, as its name implies, to provide rest for your back when sitting. It does so by reducing the tension in the back muscles, tendons, and ligaments, and the compression in your lumbar spine. The backrest also has the side benefit of reducing the pressure on your sitting bones. The effectiveness of a backrest depends on its inclination, angle with the seat pan (**backrest angle**), dimensions, contours, degree of contact with the back, presence of a lumbar support, and adjustability.

When you sit with your back upright without a backrest, the weight of your upper body rests mainly on your spine and buttocks. The weight is pulled downward along the spine toward the seat. This weight is responsible for the compression on the sitting bones and spinal discs. The pressure is greatest in the lumbar (lower back) discs of the spine. When leaning on a backrest, however, part of the upper-body weight is transferred to the backrest, lessening the weight transferred downward along the spine. This reduces the pressures in the lumbar discs and buttocks. The further back you lean on the backrest, the greater the amount of upper-body weight transferred to the backrest. The amount transferred down the spine and the pressure on the spinal discs and sitting bones will then be smaller.

Leaning on a backrest also reduces the pull tension in back muscles, tendons, and ligaments, because your trunk is stabilized and there is no need for any strong back muscle contraction for maintaining balance. Nor is there much need for the ligaments to contract strongly, because the joints in the spine or other parts of the trunk are not sharply bent. Leaning back beyond a certain point, however, offers little further advantage in reducing the mechanical stress on your back. Experiments have shown that lumbar disc pressure is greatest when you lean forward on a horizontal seat and that it decreases sharply as you lean backward. However, as you lean beyond about 20–30 degrees from the vertical, the benefit decreases. Leaning backward about 10–20 degrees from the vertical (with a hip angle of 100–110 degrees) when sitting on a horizontal seat is adequate for prolonged work.

The HFES recommends that the angle between the backrest and seat pan should permit you to sit with a hip angle of at least 100 degrees. You should, therefore, make sure that your backrest can tilt beyond 100 degrees.

Leaning back sharply while sitting and working may be beneficial to your health, but it affects your work efficacy. You lose some control of your hands at the keyboard, the strain in your shoulder muscles may increase, reading from the computer screen may become difficult, and there may be a slight tendency to slide forward on the seat.

Hand control is lost because your arms move away from their positions over the keys as your trunk and shoulders move backward. You must then reach forward to place your fingers over the keys. When you reach forward, your shoulder muscles must work harder to hold your arm suspended in space. Of course, you could use a forearm pad or wrist pad in front of the keyboard to alleviate the strain from your shoulder muscles, but this may decrease your speed of keying. If your work does not require great speed of keying, a wrist or forearm rest pad may be helpful.

As your upper body moves backward to make contact with the tilted backrest, your line of sight becomes elevated and you are forced to bend your neck sharply forward to keep your eyes on the screen, making your neck muscles work harder. Your buttocks may also tend to slide forward when you lean back sharply, and the muscles in your legs and trunk must then contract more strongly to plant your feet more firmly on the floor.

It is largely because of these disadvantages that many operators prefer not to lean backward sharply, even though they are aware of its benefits. A backward tilt between 100 and 110 degrees is best for computer operators. It is a range that gives relief from musculoskeletal strain in your back in addition to allowing you to maintain hand control. Some people may prefer a backward tilt greater than 110 degrees.

You can determine your best seat-backrest angle by trial and error. Experiment with your chair. Change the seat-backrest angle a few times during the day or over several days if necessary. After a time you will find the position most suitable for you. It is likely that this range will be within or close to 95–110 degrees. If it is just outside of this range, you shouldn't worry. The most suitable range is that which eliminates musculoskeletal fatigue, or delays it as long as possible, and allows you to work efficiently at the same time. Do not deliberately set the angle on your chair at a specific value between 95 and 110 degrees. Remember that these may be suitable for the majority of people, but you may not be in that majority.

ADJUSTABLE SEAT PAN AND BACKREST ANGLES

The backrest and seat pan of a chair should not be treated as separate entities. They are interrelated. Together, they determine your sitting posture at work and, as a result, the amount of mechanical pressure your lower back and buttocks are subjected to. If the backrest angle is about 90 degrees, then you will be sitting in a right-angled posture. Most traditional (nonadjustable) chairs were designed to keep you sitting in this posture. In some of these chairs the seat pan is horizontal and your trunk would, therefore, be vertical. In others, the seat pan is slightly tilted upward at the front, and your trunk is slightly tilted backward. Modern adjustable office chairs no longer have a fixed seat pan and backrest. Many have an angle-adjustable backrest with a fixed seat pan tilted about five degrees upward. Others have both an angle-adjustable backrest and seat pan. As mentioned before, the five degree slope in fixed seat pans is intended to prevent sliding forward on the seat. An adjustable seat pan allows you to either raise the front and lower the rear or lower the front and raise the rear to achieve a wide range of hip angles and sitting postures.

Designing chairs so that both the backrest and seat pan can be adjusted was a revolutionary step based on two ergonomic and orthopedic ideas. One is that a more open hip angle (greater than 90 degrees) imparts less mechanical pressure in the lumbar spine than a right angle, as explained earlier. An open hip angle helps preserve your natural lumbar curvature. The other concept is that changes in posture help to relieve some of the strain in over-stressed muscles and joints by redistributing the strain of your upper-body weight and that posture can easily be changed by changing the geometric configuration of the chair.

Of all these different designs, a chair that has both an adjustable backrest and seat pan is the best because you can set not only the angle at your hip but also the position of your trunk and thighs to achieve many different sitting postures. For example, you can achieve a 105 degree hip angle with either a horizontal or a downward tilted seat pan or with a vertical or inclined backrest. The chair with the fixed seat pan and adjustable backrest will not allow you to maintain an upright trunk when the hip angle is opened widely, but it does allow some amount of adjustability in sitting posture.

LUMBAR SUPPORT

As mentioned earlier, the pressure in your spinal discs is greater when sitting without a backrest than when standing. At the lumbar discs, it is about 35% greater when you sit upright and twice as great when you sit with your trunk bent forward. Leaning on a backrest reduces some of this pressure. But there is a special feature in the backrest that can further reduce the pressure. This is the **lumbar support** or **lumbar pad**. Have you ever placed a small pad or pillow against the backrest of an ordinary chair in the area of the lower back to relieve pain or discomfort? It works. There is firm scientific support for this makeshift lumbar support.

Remember that sitting tends to flatten the forward curve in the lumbar spine, building up pressure in the lumbar discs. A lumbar support is a forward curved portion of the backrest of a chair near the bottom. When you lean firmly against the backrest, it presses forward on the lower back and resists the flattening of the lumbar curve. Once the forward curve is maintained, pressure is prevented from increasing significantly. Leaning against a lumbar support has the same effect as pressing your hand against the curve in your lower back, as many people do when they want to relieve backaches after prolonged sitting. The effectiveness of the support depends on how well it fits your lumbar curve. It should not be placed too high on the back, above the lumbar curve, or too low, below it. Nor can its curvature be too slight or too sharp.

More recent ergonomically designed chairs allow you to vary the height of the backrest to reposition this lumbar support appropriately. Some chairs also allow you to vary the curvature of the lumbar support. Backrest height is usually varied by manipulating a knob located near the bottom of the backrest. A tall person with a long trunk can raise the backrest, and a very short person can lower it to position the lumbar support in the correct place in the lower back. In some sophisticated ergonomic chairs, the lumbar support is a bag that can be inflated with air or deflated. The more air that is pumped into it, the greater the lumbar curvature becomes and vice versa. This inflatable type of lumbar support has been employed in the driver's seat of some cars for many years now. Long hours of driving without a lumbar support are likely to produce fatigue and pains in the lower spine. Pains can be distracting enough to cause drivers to make critical errors that can lead to accidents. Lower back fatigue while working at the computer workstation can be distracting enough to reduce your work efficiency and productivity.

If your chair has no pronounced lumbar support, you can attach a lumbar pad at the lower back area of the backrest. A variety of lumbar pads are available in computer and office furniture retail stores. In your home, where you may not be concerned with office aesthetics, you may use a small, firm pillow instead.

STRESS IN THE SHOULDERS AND BACK

Armrests may be unnecessary when using a keyboard or a mouse but are useful during short work pauses. When the arms hang downward, the shoulder ligaments bear their weight, but when they are extended forward and suspended in midair, as when using a keyboard that is too far away, the shoulder muscles and tendons bear their weight and the strain is much greater. The total weight of your arms is 10% of your body weight. So, if you weigh 140 pounds, your shoulder muscles must support 14 pounds. This is a tremendous burden to bear for prolonged periods of work. Resting your arms on an armrest eliminates most of this burden and gives your fatigued shoulders some relief.

Without armrests, the weight of your arms is transmitted down your spine and contributes to the pressure in the discs and buttocks when you are sitting. However, when you rest your forearms on armrests, the weight of your arms is no longer transmitted down your spine, and these pressures are reduced. This reduction is often welcome by people who have been using a keyboard for long periods, especially towards the end of the workday.

Armrests can be a disadvantage if your workstation is not properly designed. They may be too high, too low, too widely separated, or too close together. They should be positioned just below and outside your elbows to avoid jamming your arms when you are using the keyboard. If they are too low, you will be forced to lean to one side and use only one arm at a time. If they are too far apart you will be forced to spread your arms sideways. And, if they are too close together, they will obstruct your movement into or out of the chair, jam your arms, and squeeze you when you are sitting. Armrests may prevent you from sliding your chair under the desk or pulling out the keyboard rest even when they are the correct height.

Whether armrests are too high, too low, too far apart, or too close together depends on your body size compared to the size of the chair. In some chair designs, the armrests are adjustable. They can be moved up and down to change their height, or in and out sideways to change the distance between them. Such armrests accommodate both short and tall and thin and broad operators. They are suitable wherever different size operators share a single chair.

Some types of armrests can also rotate horizontally on a pivot so that you can position them at an angle pointing toward the center of the keyboard. Some obese or broad-shouldered people may like these kinds of armrests, because their elbows tend to be so far apart that they must turn their forearms more sharply inward than other people do.

If you are the only person who will use a particular chair, an adjustable armrest may not be necessary. But, before you install a chair at your workplace, you should ensure that the armrests are appropriate.

THE BALANS CHAIR

A strange looking type of chair has stirred much interest over the last decade. It is called the Balans chair (see Figure 9). This chair is unlike ordinary chairs in that you kneel and sit on it at the same time, and it has no backrest. The diagram on the next page shows your posture on this chair. Both the knee support and the seat are cushioned. The downward sloping seat and the upward sloping knee support allow you to sit upright with a hip angle of about 135 degrees. Your pelvis is not rotated backward and the hamstring muscles at the back of your thighs are not tightly pulled. So there are no strong forces acting on the lumbar spinal curve to flatten it. The lumbar curve maintains its natural curvature, as when you stand, and the accumulation of excess pressure in the lumbar discs is prevented. This is the real value of this chair. However, there are two serious disadvantages—pressure is put on your knees, and your legs become fixed in one position.

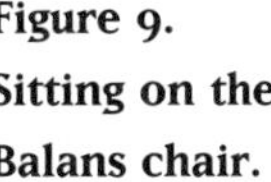
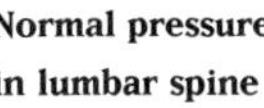
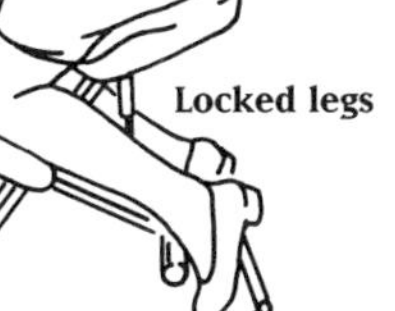

Figure 9. Sitting on the Balans chair.

People who have used this chair have declared that it is great in the beginning. They feel no undue strain in the spine and they use their keyboard normally. But, after half an hour to an hour, they begin to feel the effects of excess pressure on their knees and are forced to change to a normal office chair. When you sit on this chair, the weight of your thighs and some of your upper-body weight is transferred to your knees and, even though the knee support is cushioned, pressure builds up. In addition, your legs are locked in a flexed position, which impedes free flow of blood. This chair seems to be suitable for only about half an hour to an hour of continuous work.

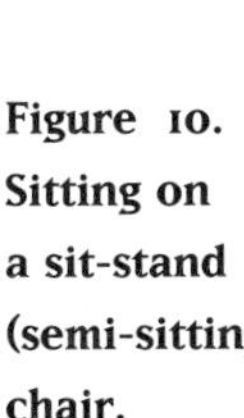

Figure 10. Sitting on a sit-stand (semi-sitting) chair.

THE SIT-STAND CHAIR

Another chair that can be useful for short periods of work at the computer is the sit-stand chair (see Figure 10). This is essentially a seat on a high base. Some designs resemble an enlarged bicycle seat on a long pole. The height can easily be adjusted. This chair is suitable for work at a high desk where you do not want to sit or stand. Because you can get into and out of it easily, this chair is also use-

Figure 11.
Working at a desk that is too low can result in (i) increased pressure on the lumbar spine and buttocks and decreased stability, (ii) increased lumbar pressure, (iii) and shoulder strain.

ful in work environments in which you must move about frequently or work in short spurts. It was originally designed for industrial work environments, but it can also be used for short spells of work with a computer at a high desk.

LEG SPACE UNDER THE DESK

The posture of your body when sitting is influenced by the amount and shape of the space below your desk. Ideally, the space should be large enough to accommodate both legs. It should be deep enough to allow you to stretch your legs out completely, high enough to prevent your knees and thighs from touching its underside, and broad enough to allow sideways movements of your knees and feet.

Limited height under the desk occurs when the desk is too low, the desk top is too thick, the drawer below the desk top is too thick, or your seat is too high. In any one of these cases, you are likely to experience difficulties in getting your thighs under the desk. You are forced to sit further away from the desk than normal to prevent your thighs from being squeezed by the underside of the desk top. The thicker your thighs, the more difficulty you experience, and the further away from the desk you must sit. To operate the keyboard you are forced to sit in one of three postures (see Figures 11 and 12):

- Reaching forward with the arms stretched outward—this can lead to shoulder strain because the shoulder muscles are forced to maintain static contraction as long as your arms are stretched out above the keyboard.

- Bending your trunk forward—this can lead to lower back strain because your back muscles are under static contraction.
- A combination of the above two postures—this can affect both your shoulders and your lower back.

Many desks contain a drawer under the top surface exactly where the thighs must be placed. Though these drawers are useful, they rob you of space for your thighs. When buying or selecting a desk, you should not be carried away by the volume of storage space in the top drawer. Always consider that an extra centimeter of thickness in this drawer may be responsible for an awkward body posture and consequent aches and pains. Some computer desks are designed with a roll-out keyboard rest under the desk top. This convenient feature also limits the height under the desk available for your legs. A more recent and suitable desk design comes with a keyboard rest with adjustable height and tilt.

If you want to buy a special desk for your computer workstation, you should determine if there is enough space under it by sitting in front of it with the keyboard and screen in place. It is often difficult to make this judgment by considering the desk in isolation, and it is impractical to do so by looking at a picture of a desk in a sales magazine.

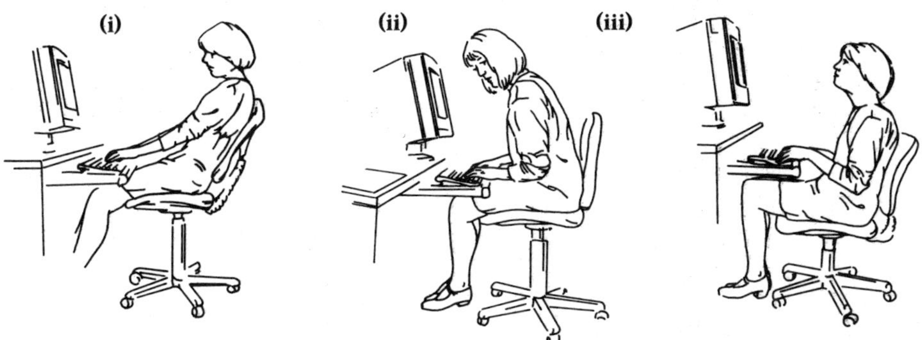

Figure 12.

Other constrained postures at the computer workstation can result in (i) increased disc pressure and shoulder strain, (ii) a twisted trunk and neck strain, (iii) and neck and wrist strain with increased pressure on the buttocks.

While sitting and working, there are times when you need to stretch your legs under your desk to ward off the feelings of fatigue. Extending and flexing your knees alternately allows your leg muscles to contract and relax rhythmically and promotes blood circulation. If you are very tall, you have to be extra careful that the space under your desk is deep enough to accommodate the length of your legs. If it's not, you will have to move your chair backward. This is undesirable because it increases the distance between your body and the keyboard or screen and forces you to extend your arms or bend your upper body forward. The space under the desk should also be wide enough to allow you to move your legs sideways. Do not store objects, such as boxes, purses, or briefcases under your desk. It robs you of valuable space and forces you into these constrained postures.

Low desks make it difficult for you to get your thighs underneath them. Simply lowering your seat will not help if you are tall. A very low seat produces unique problems. First, your buttocks will fall below the level of your knees, making your thighs slope upward towards your knees. (An exaggerated example is an average adult sitting on a chair made for a kindergartener.) Your upper thighs will not make firm contact with your seat, and your body will pivot on your sitting bones. The pressure on your buttocks will increase. And you will be unable to lean forward on your thighs to relieve buttocks pressure, as many people do when using a seat at normal height. Second, your hip angle is reduced to less than 90 degrees, which causes the hamstrings to pull more strongly on the pelvis. Recall that the pull force is transmitted to the lumbar spine and flattens the lumbar curve, increasing the pressure in the lumbar discs. Third, free movement of the legs is inhibited because of the limited space under the low seat pan. This prevents you from tucking your feet under the seat to get relief from fatigue in the lower back and legs.

Chair Base and Casters

Casters are the wheels on the base of a chair. They help you roll the chair easily around your workstation. You can purchase a chair with casters or add them separately. They are especially useful when you are working with several documents or other materials spread over a wide area at your workstation.

To promote free rolling of the chair, the floors of workstations are usually covered with hard plastic floor mats. However, movement on these kinds of mats may be too free for some people. A slight push with your legs can propel you backward. To prevent this, opposing muscular work by the legs is required. The effects of this extra work over several hours can lead to leg fatigue. If your work area is small and your chair moves unnecessarily because it stands on one of these mats, you may remove the mat and allow the casters to rest on the carpet or refit your chair with special casters that do not roll as easily.

You have a few choices in the type of casters available. In one kind, the wheels can be locked by pushing a small lever next to your feet. For carpeted floors, you can purchase casters made of hard plastic, and for hard tiled floors, you can purchase casters made of soft plastic. There are also design features in casters for preventing a chair from rolling away. Some casters lock when you sit on the chair, allowing you to roll the chair only when you stand up. This is useful for slanted or slippery floors. Another kind of caster locks when you rise from your seat, so your chair remains in position for you to sit down again.

If you look at the base of a modern office chair you will notice that there are five legs instead of the traditional four. This is to provide more stability and decrease your chances of toppling over when you lean backward.

Footrests

A footrest at a workstation is an object that compensates for either short legs or a high seat. Its purpose is not to rest your legs after they have become fatigued but to prevent foot and leg fatigue from occurring in the first place. However, there seems to be a misconception about footrests in computer workstations. Some people believe that

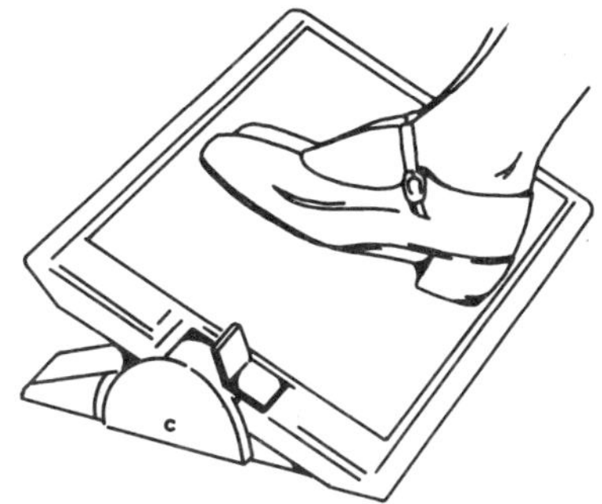

Figure 13. Footrest with a tiltable surface.

their workstation is incomplete without a footrest. Most of these people are using footrests unnecessarily. Our susceptibility to advertising is partially responsible for this misconception. Advertisements for computer furniture almost always include footrests, and some buyers do not question their necessity.

The use of a footrest in the office implies that there is a mismatch between your body size and the size or arrangement of the furniture at your workstation—lack of proper adjustment of the heights of your desk and chair may prevent your feet from resting comfortably on the floor. Common problems that prevent your feet from resting flat and comfortably on the floor include tall desks, fixed-height chairs, people who are short, or a combination of these. As you probably realize by now, tall desks and fixed seat heights are considered poor features of computer-furniture design. Fortunately, these can be corrected by using ergonomically designed chairs and desks. If you cannot afford a height-adjustable desk, then a height-adjustable seat may solve the problem—as long as the desk is not too high. But, if you are very short, you may find it difficult to place your feet flat on the floor even when your seat height has been adjusted downward. In this case you need a footrest.

There are some other important things you should know about footrests before selecting one:

- It should be adjustable enough to allow you to move your feet at the ankle. This means that it should be rotatable in the vertical plane to allow you to flex and extend your ankle joint without taking your feet off of it.
- If its surface is not adjustable, then it should be angled upward about 20 degrees toward your toes to allow your ankle to maintain a comfortable angle. The large foot pedals of the old manual sewing machines were angled in this way, as is your car's gas pedal. There are some footrests on the market that can be tilted upward to almost 25 degrees.
- It should be broad enough to accommodate both feet and allow them some sideways movement.

  It should be deep (long) enough to allow some stretching of the legs, but not too long so that it becomes an obstacle on the floor. Remember that a footrest is an extra object on the floor that should be avoided if possible.
- Its surface should have enough frictional resistance or your

feet can slide, requiring extra muscular work in the legs to prevent the sliding.

Avoid chairs or stools with a foot ring around the base. They are ineffective for prolonged sitting and unsafe as footrests, unless your shoes have no heels. They are ineffective because they constrain the legs to a specific position, with the knees flexed and the feet tucked under the seat. Heels on your shoes can easily hook onto the foot ring when you try to move your legs, causing you to fall. People often make mistakes that seem to require only common sense when they are distracted by more important things at work. Not taking your heel off a base ring is one such mistake that many people have made. Employers should take note of this, since they bear responsibility for safety in the workplace. They can claim that a worker who falls did not use common sense in getting off the stool, but the worker can also claim that the employer provided her with unsafe working equipment.

Palm Rest or Wrist Rest

A palm rest, or wrist rest, is firm padding that can be placed snugly in front of the keyboard. You can rest your palm or wrist on it while keying. The main benefit of resting the arms in this way instead of suspending them in midair is that you reduce the strain in the shoulder muscles. This type of rest is suitable when you must wait on the computer to respond with messages or other prompts. However, it may take some time for you to get accustomed to it. Some people, especially speed typists, may not like a palm or wrist rest because it may obstruct free finger movement.

The following should be considered when deciding whether or not to use a palm rest or wrist rest:

- It should be soft with a rounded front edge.
- Its height should allow you to maintain your wrist at its natural angle while keying. You should not be forced to bend the wrist downward or upward.
- It may be unnecessary when using a mouse since your hand must move about the mouse pad.

Document Holders

The use of document holders for positioning documents is quite common in the computer workplace. Document holders are more than a passing fad, they are excellent ergonomic aids. They have useful orthopedic values and improve productivity. To appreciate the use of a document holder, you must consider the difficulties of working without one.

Most computer work is done with documents that cannot stand up by themselves. If you don't have a holder, the document must be placed flat on the desk. You are then forced to bend your neck downwards to be able to read from it. In some cases, you may even bend your trunk forward. If the writing on the documents is too small or not easily legible, you will bend even more sharply to bring your eyes closer to the documents. These postures place the neck and lower back under considerable mechanical stress, and aches and pains are inevitable.

The way to avoid all of this is to place the document on a document holder, which is positioned upright. The need to bend over will diminish, allowing you to hold your head in a more comfortable position. If the print is difficult to read, you can look at it more closely by moving your chair or the document holder forward. You can find document holders that rotate vertically and laterally in office supply stores. Some newer computer furniture designs have the document holder on a pivot arm so that you can position the document between the screen and the keyboard below screen level.

Adjustable Furniture

We have seen that the modern office chair has evolved into a complex machine, utilizing complex technology. We cannot assume that everyone who sits on a modern office chair knows how to use all its features. Managers and supervisors of computer workplaces will find it beneficial to educate their workers in the value of adjusting their chairs and desks. Of course, they must first educate themselves. If you know how to make adjustments, you will be more likely to make them. The adjustments are simple and quick to make. With a little practice, you should be able to make them in less than a minute. From a management point of view, this can be regarded as an exercise in health and safety management.

Many office workers do not make the necessary adjustments because they have never done so and assume that the adjustments are difficult or time consuming. Ergonomically designed furniture is getting progressively easier to adjust. There is fierce competition among manufacturers for selling ergonomically designed furniture, and they realize that the acceptability of their products depends on their ease of use. Most of the adjustment mechanisms on modern ergonomically designed chairs are located under the seat, at the rear, or on the sides of the seat pan. In most cases, you can make the adjustments while sitting, with little physical effort. A simple demonstration by someone on how to adjust chair features will take only a few minutes. Special tools, such as wrenches and screwdrivers, are not required in a well-designed chair, and there is little chance that your hands would become covered with grease.

Bear in mind that small, seemingly trivial adjustments to your chair can produce great relief from the amount of work the muscles, joints, and other structures in the body are required to perform. Supervisors and managers should also consider this seriously. Adjusting a chair or desk is often as important as adjusting the brightness of the screen, the overhead lighting, or any other aspect of computer work.

In today's workplace people are often under pressure from management to maintain a high rate of production. Some people may consider the time taken to adjust seats as wasteful. But they should realize that trying to maintain a high production rate under the burden of aches and pains can easily lead to cumulative trauma disorders as well as a decline in productivity. The effects can be devastating to your health, as described earlier in this book. A few minutes taken to adjust the chair will be beneficial to both the worker and the employer. Reducing or eliminating musculoskeletal stresses and strains can translate into real dollar savings from less health maintenance and greater productivity. Reduction in medical costs, sick leave, workers' compensation benefits, and insurance premiums are real results of providing workers with well-designed computer workstations. From a productivity point of view, there can be less absenteeism from cumulative strain illnesses, reduction in errors, and increased output; from a psychosocial point of view, there can be increased job satisfaction and higher morale among workers, less lost time, and less worker turnover.

Successes in Workplace Design

The use of ergonomically designed furniture and other aspects of ergonomic improvement in the office and other computer workplaces has been remarkably successful. They have not eliminated aches and pains in the workplace (only a dreamer would expect that), but they have yielded significant improvements in comfort, enhanced performance, and reductions in health complaints. There are numerous examples of these success stories, and a few are mentioned here as highlights.

At one airline computer center in Singapore in the 1980s, musculoskeletal and visual problems were rampant; 94% of data entry operators and 77% of conversational operators had visual problems, stemming partly from poor illumination. Workplace-design improvements were then applied. They included the use of ergonomic furniture, footrests, and document holders, changes in illumination, and more frequent rest breaks. After a year, musculoskeletal problems were reduced to half their original rates and visual problems to two-thirds. In addition, keying speed increased by 37%, and error rate dropped 93%. Many other studies around the world have reported productivity improvements of 15–25%. Laboratory studies at the National Institute of Occupational Safety and Health (NIOSH) have shown a possible increase in performance of 25% with improved furniture and workplace design and 18% with improved lighting conditions.

Of course, the degree of improvement depends on the conditions of the workplace prior to the implementation of ergonomic changes. Spectacular improvements are unlikely to occur in a computer workplace that is already well designed. These figures should, however, serve as a reminder to workers and employers about what good furniture and workplace design can achieve.

POINTS TO REMEMBER

- To avoid aches and pains and maintain your work productivity in the computer workplace, use ergonomically designed furniture and arrange it properly.
- Use adjustable furniture at your computer workstation, especially if more than one person will use the workstation.
- The chair is the most important piece of furniture in computer work, but it must be compatible with the design of the desk and the nature of the work.
- The backrest should be broad enough to support your entire back, but not too high to impede neck movement or too broad to impede elbow movement.
- The chair base should have five legs for stability.
- Footrests should be used only when faults in other workplace-design features make them necessary.
- The front edge of the seat of your chair should be curved downward to prevent pressure against the back of your knees.
- A desk that has limited leg space will restrict leg movements and may force you to sit in contorted postures.
- Avoid resting your wrists on the hard edge of the desk or keyboard stand.
- Try using a pad on the desk edge or a palm rest in front of the keyboard.
- Use a document holder to avoid bending your neck or trunk downward sharply.
- When seated at your computer, your upper arms should be approximately alongside your trunk, and your forearms should be horizontal or inclined slightly upward. Adjust the height of your keyboard, desk, seat, or all three to achieve this posture.
- Your feet should rest flat and comfortably on the floor.

CHAPTER 12

REPETITIVE MOTION INJURIES

One of the most feared types of illness in the modern workplace occurs in the hand and wrist. It is called **carpal tunnel syndrome**, after the part of the body (the carpal tunnel in the wrist) that is affected. But carpal tunnel syndrome, or CTS, is really one of a set of related occupational illnesses arising from the accumulation of traumas from excessive motion when one specific part of the body repeats a certain motion thousands of times a day. An example of this in computer jobs is the fingers using a keyboard. The fingers get little rest, and these small, imperceptible, and otherwise harmless strains gradually build up each day. After several months or years, they wear away the tissues in the hand, producing inflammation of the surrounding areas and causing other problems with nearby nerves. The symptoms are tingling sensations, numbness, swelling, pain, restricted motion, and loss of hand strength. Such hand disorders are popularly known as **repetitive motion injuries (RMIs) or repetitive strain injuries (RSIs).**

RMIs are part of a larger set of work-related disorders that encompass illnesses arising from the cumulative effects of sustained strains. These strains include the tensed muscles, tendons, or ligaments when your body maintains a constrained posture for a long time. In the United States these health problems are together referred to as **cumulative trauma disorders (CTDs).** This chapter discusses the CTDs associated with RMIs, not those associated with sustained static contractions of muscles or ligaments.

In computer work, RMIs are caused by rapid and repetitive motions of the fingers over a prolonged period of time. If you perform data entry or word processing tasks, you are at great risk of developing these disorders because of the tremendous amount of work your fingers perform each day. A keying rate of 17,000 strokes per hour is common in data entry jobs. This translates into approximately five keystrokes per second or twenty-five and one half million per year (assuming that you enter data for six hours per day, five days per week, fifty weeks per year)—an incredibly high workload. Many skilled typists performing word processing can achieve typing speeds of over ninety words per minute with a typical computer electronic keyboard. Given that the English language has an average word length of five letters, this speed translates into seven to eight finger motions (keystrokes) per second, or thirty-two million per year!

Complaints of arm-hand problems are so numerous among computer operators that these occupational illnesses have been called the "epidemic of the nineties" by the news media. Although this may be an exaggeration, the complaints are real, and the illnesses are so widespread and debilitating that they warrant serious investigation. RMIs do not occur in isolation in the body because the hands do not work in isolation. If you suffer from RMIs in the hands and wrists from computer work, you are likely to suffer from other strains resulting from prolonged static muscular contractions. These include headaches and aches in the neck and lower back.

Changes in Office Work

Before computer jobs arrived in the office, traditional typists performed a variety of tasks. These gave them the opportunity to move various parts of their bodies, preventing severe accumulation of strains in the hands, neck, or back. They answered phones, filed papers manually in cabinets, changed papers on the typewriter, applied liquid eraser to documents, and took frequent short pauses. They also had the opportunity to talk to coworkers and move their heads and arms. During these periods, keying ceased, the fingers were relieved of their tension, and other muscles of the body were exercised. These motions produced rejuvenating effects on the body similar to the exercises that computer workers are encouraged to perform.

In today's computer office, jobs are highly specialized, and you seldom have the opportunity to vary your body movements. Some jobs are reduced to one simple action, requiring repeated use of a few muscles in the body throughout the workday. When you look at experienced typists or data entry operators using a keyboard, you cannot fail to notice how still their bodies remain while they work. For long periods of time, only their fingers seem to be alive. Their hands are kept in position over the keyboard and their fingertips are positioned over specific keys, depressing them with brisk downward strokes. The fingers work like small pistons, pivoting about the knuckles. Data entry jobs require you to tap on the keys with your fingers all day. The tension in the finger tendons from one keystroke may not be great, but the gradual accumulation of the effects from millions of keystrokes is likely to overwhelm your body's recuperative powers.

When electronic keyboards were invented, many people thought that keying tasks would have become easier. But the increasing number of health complaints in the hands and wrists by computer operators indicate clearly that other important factors, apart from finger force for keying, determine the level of stress in computer work. These other factors include the repetitiveness of finger actions, the long periods of uninterrupted work, inadequate rest, and constrained arm positions. Muscular contractions that move the fingers generate the forces required for depressing the keys and returning the fingers to their original positions. These forces are a small percentage of the maximum strength of the fingers, but they are repeated so quickly and for such extended periods of time that the hand easily becomes fatigued from the strain.

Another misleading idea is that the period between workdays (for example, from 5:00 P.M. one day to 8:00 A.M. the next day) is long enough for the hands to recuperate from the strains of repetitive muscular work. However, many affected computer operators are often disappointed because their aches and pains return soon after they start working the next day. A night's rest may not be adequate for full recovery, and the next day's work simply adds to the residual effects of the previous day's work. Eventually, the pains may become chronic. As the months and years pass, the gradual accumulation of strains in the tendons may cause inflammation and other deteriorating conditions.

Types of Arm-Hand Disorders

The most commor arm-hand occupational disorders from prolonged, repetitive keying include **tendinitis, tenosynovitis,** and **carpal tunnel syndrome**. Others, such as **epicondylitis** and **bursitis**, are less common. These terms are often confused with **repetitive strain injury (RSI), repetitive motion injury (RMI),** and **cumulative trauma disorder (CTD).** The confusion arises because of the perspective from which the disorders are viewed. They are named according to either the part of the body damaged and the nature of the damage or the nature of the job that is believed to have caused them. Tendinitis, tenosynovitis, carpal tunnel syndrome, and bursitis are named according to the nature of the disorder, while RSI, RMI, and CTD are named according to the characteristics of the work that is believed to cause them.

There is also considerable confusion in the diagnoses of these illnesses. Physicians have often found it difficult to determine whether the aches and pains are caused only by strained tendons, inflamed tendon sheaths, strained ligaments, inflamed bursas, or a combination of these.

TENOSYNOVITIS AND TENDINITIS

Tendinitis refers to inflammation of the tendons that connect a muscle to a bone. Tenosynovitis refers to inflammation of a tendon within its protective sheath or inflammation of the sheath itself. Tendons can be damaged from repeated tension in the same way a tightly pulled rope can fray and tear.

There are many regions in the body, especially near joints, where a tendon runs next to a hard or tough surface, such as a bony projection or tough connective tissue. This makes the tendons susceptible to injury because they rub against the tough surface when the joint bends sharply or when the muscles of those tendons contract. But nature has anticipated tendon damage in your body and has taken steps to protect the tendons from frictional wear and tear. A protective covering called a **tendon sheath** surrounds the tendons near the dangerous regions.

The tendon sheath is a tube surrounding the tendon cord where it passes next to a tough or hard surface. It is made of tough connective tissue and is filled with a fluid called **synovial fluid**. When a joint bends, the sheath acts as a barrier preventing direct contact of the tendon with the adjacent bone or tough ligament. But the sheath itself may rub against the hard surface. If the rubbing is neither intense nor prolonged, the sheath will not be damaged, and the tendon within it will be perfectly protected. The synovial fluid provides lubrication to help the tendon glide smoothly within the sheath when finger muscles contract and pull their tendons. This fluid also provides nutrition to the tendon and its sheath. In normal activities of daily living and in most types of jobs, the protection provided by the synovial sheath is adequate. In computer work, however, the tremendous amount of finger motions may cause mechanical wear and tear, rendering the finger tendons susceptible to inflammation.

Friction from excessive motions also breaks down the protective properties of the synovial fluid. It produces heat around the area where a tendon rubs against a harder surface and causes the lubricating fluid to become less viscous. It is like your hot car engine causing the lubricating engine oil to become thinner. A dangerous cycle is then set in motion. A thinner lubricant for the tendon leads to increased friction between the tendon and its sheath, greater heat production, greater loss of fluid viscosity, and faster wear and tear on the tissues. Eventually, the tendon or tendon sheath becomes inflamed. This inflammation is the body's reaction for protecting itself from subsequent invasion by bacteria.

In computer work, both tenosynovitis and tendinitis are believed to be caused by the repetitive motions of the fingers while using the keyboard, especially when you work with your wrist bent appreciably. As mentioned earlier, some types of computer jobs may require as much as seven downward and upward motions of the fingers per second. The muscles in those fingers contract and tense with each motion. If the wrist is bent upward, the tendons rub against the wrist bones; if it is bent downward, the tendons rub against the transverse carpal ligament. Nature did not provide protection against the rigors of modern day computer work. An anatomist might say that the human body was not designed for computer work. An anthropologist might say that changes in methods of manual work are evolving far faster than changes in the human body and mind can keep up with them. An ergonomist would agree with both of them and would recommend that the method of work and the equipment be redesigned to conform to the functional capacities of the working population.

CARPAL TUNNEL SYNDROME

Carpal tunnel syndrome is the condition in which the median nerve (one of the nerves controlling the fingers) is compressed inside the wrist. Repetitive tension and friction of the tendons in the wrist may cause them to become swollen and press the median nerve against the wall of the wrist. The following description of the anatomical structure of the hand and wrist illustrates how this occurs (see Figure 14).

The fingers are moved by muscles whose large mass lies in the forearm just below the elbow. The cordlike tendons at the elbow end of these muscles are attached to the bones in the elbow. Some of the muscles do not have tendons at the elbow end and attach to the elbow bones directly. The tendons at the other end of the muscles run down the lower forearm, traverse the wrist, and run along the fingers where they attach to the **phalanges** (finger bones).

In addition to the tendons of the finger muscles, blood vessels and nerves that control the action of the muscles also run along the forearm toward the fingers. In their passage from the forearm to the hand, the tendons and the median nerve traverse a narrow channel in the wrist called the **carpal tunnel**. This tunnel is just about the size of your thumb, and just large enough to accommodate the structures passing through it. The tendons must bunch together in the wrist in order to pass through this choke point. The inner wall of the carpal tunnel is not soft. It is composed of

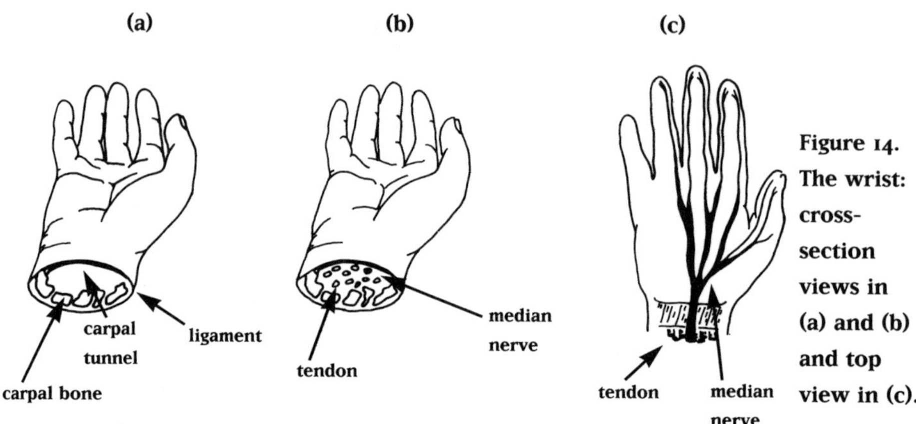

Figure 14. The wrist: cross-section views in (a) and (b) and top view in (c).

tough connective tissue called the **transverse carpal ligament** at the bottom of the wrist and an arch of seven small bones, the **carpal bones**, at the sides and top of the wrist. The movement of your fingers during computer work requires contractions of muscles and consequent tension in their tendons. When you work with a sharply bent wrist, you put greater strain on the finger tendons than with a straight wrist. A bent wrist weakens the force capabilities of your finger muscles. These muscles must therefore work harder to move the fingers. This means that friction in the carpal walls will increase. So maintaining this kind of finger work for many hours each day places tremendous mechanical strain in the tendons, and wear and tear is bound to occur. Sharp bending of the wrist also reduces the size of the carpal tunnel like a bent water hose and increases the pressure on the tendons and the median nerve. Even if the wrist is not bent sharply and the rubbing on the carpal wall is not strong, the repetitiveness of the finger motions during computer work is likely to wear the tendons or their sheaths. Each keystroke involves tension in the tendons and some rubbing. Even though one stroke is harmless, the accumulation of thousands per day can easily overwhelm the recuperative capacities of the tendons.

Inflamed tendons become swollen, occupy more space inside the carpal tunnel, and press on adjacent structures. One of these compressed structures is the median nerve. When it is pressed against the walls of the carpal tunnel, the result is not only pain but also some loss in function of the nerve. In addition, as carpal tunnel syndrome progresses, the regions of the hands and fingers connected to the branches of the median nerve experience numbness, handgrip and finger-pinch strength weaken, and some amount of hand and finger control is lost. The affected regions are the thumb, index finger, middle finger, half of the ring finger, the palm of the hand, and the large muscle mass at the base of the thumb. The loss of control and strength in the hand occurs mainly because the muscles of the thumb are affected. As inflammation worsens, the tendon sheaths become more swollen as they overproduce fluids, and the connective tissue within the carpal tunnel proliferates as the body begins to fight against the inflammation. At this point, you would hardly be able to perform even common household chores, and life would be unbearable without proper medical treatment.

People suffering from carpal tunnel syndrome may also complain of pain along the arms and shoulders. This is because the affected median nerve originates in the neck and runs along the arm toward the fingers. Three main nerves—the ulnar, median, and radial nerves—control the muscles of the hands and fingers. They branch off from the spinal cord in the region near the base of the neck, then run along the arm to the fingers. The ulnar nerve runs near the ulna bone in the forearm and terminates in the little finger. The radial nerve runs near the radius bone of the forearm and terminates in and around the fingers near the thumb. The median nerve takes a middle course through the carpal tunnel and terminates in the palm of the hand, the first three fingers, and half of the ring finger.

If you have ever suffered from a whiplash injury to your neck, you may experience the symptoms of CTS from computer work. In a whiplash injury, the neck vertebrae may be displaced and press on the nerves that branch from the spinal cord in the neck. Bending the weakened neck when working at the computer can then cause the vertebrae to press on the nerve branches in the neck. The resultant pain may then be felt not only in the neck but also along the path of the nerve(s) in the arm, including the wrist. It is also possible that prolonged sharp bent neck posture while working can cause pressure on the nerves at the neck region, even without previous neck injury.

Report any of the above symptoms to your employer as soon

as you first notice them and request an ergonomic evaluation of your workplace. Remember that according to OSHA, your employer must provide a safe and healthful workplace for you. You can save your employer money and keep yourself healthy by reporting stressful work conditions promptly. An enlightened supervisor will appreciate your input in making the workplace healthier.

In its early stages, CTS can be reversed if you cease doing the job that caused it and seek medical treatment. Since giving up your job is not practical, you must seek to improve your work conditions. It is widely known that many workers do not report symptoms of CTS to their employers for fear of being considered poor performers or problem workers, either of which can lead to their being laid off later. Many affected workers are also not aware that the symptoms of CTS are directly related to their work. They often wrongly blame themselves for being too weak or unfit.

At work, two main conditions may cause you to bend your wrist: the design of the keyboard and the design and arrangement of your furniture. The bending is often sideways (toward the little finger) due to keyboard design. The bending upward can also be if the keyboard is too low or downward if the keyboard is too high. Keyboard design flaws will be discussed in the next chapter (see Figure 15).

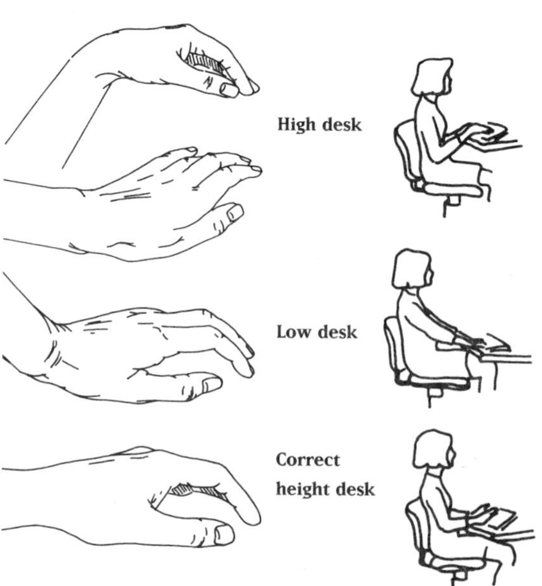

Figure 15. Different wrist positions while using a keyboard. Only the straight wrist position (BOTTOM) produces little stress.

When you use your keyboard, your hands should be in their natural position, that is, with a slight backward bend of the palms. If your seat is too high or your desk is too low, you will be forced to bend your wrist further backward to position your fingers properly over their keys. Very tall people are more at risk for this kind of mismatch than shorter people. Similarly, if your seat is too low or your desk too high, you will be forced to bend the wrist downward. Very short people and people who slouch in their chairs are at greater risk for this kind of mismatch than other people.

One reason physicians used to be reluctant to attribute CTS to any type of work was that it was known to be associated with other adverse health factors unrelated to work. They include joint problems, such as rheumatoid arthritis and malaligned fractures, and fluid retention conditions in women, including the use of oral contraceptives, pregnancy, menopause, and diabetes. Pregnant and older women operators are, therefore, at greater risk, complicating the identification of the causes of CTS. Some physicians also associate low levels of vitamins B_6 (pyridoxine), C, and E with CTS. All these factors tend to reduce the size of the carpal tunnel, and computer operators who are affected by any of them are more susceptible to CTS.

Physicians diagnose CTS by checking for decreased sensation in the hand, difficulty in flexing the wrist, and decreased ability of the median nerve to conduct impulses. At a certain stage of CTS, a physician may treat you with anti-inflammatory medication. If the disease worsens beyond a certain point, he may recommend surgery. In this procedure, an incision is made near the palm of the hand, cutting the traverse ligament so that the pressure on the median nerve is reduced. Like all other types of medical problems, you should not agree too readily to CTS surgery. There is concern in the medical profession that it is sometimes performed prematurely and unnecessarily. You should get a second medical opinion before you decide on such surgery. Moreover, you should know that surgery does not cure work-related CTSs if you return to the same job that caused it. Surgery and medicines do not make you stronger, they simply arrest a deteriorating condition and prepare your body for healing.

The use of wrist straps by people who feel pain in the wrist has become popular. Medical professionals recommend them, employers advise their workers to use them, and vendors advertise them intensely. But they may not really be useful. Some vendors claim that these wrist devices can prevent CTS. These claims have never been proven to be true by independent researchers unassociated with the manufacturers or vendors of these devices. In fact, it is doubtful whether they help at all. Wrist straps or splints may even make a bad situation worse. Their main purpose is to restrict bending of an injured wrist to enhance healing and reduce pain. But if you are forced to bend your wrist because of poor workstation arrangement, these devices will make it more difficult for you to perform your job and may create other problems. For example, if your keyboard is too high you will most likely raise your shoulders to compensate for the loss in bending the strapped wrist. This leads to shoulder strain. Moreover, to restrict bending of the wrist, these devices must be strapped firmly, and this can prevent free movement of the tendons on the palmar side of the forearm, just above the wrist. Clasp one hand about your wrist of your other arm tightly. Now try to move your fingers of the clasped hand and note how difficult it is. This is what happens when you use a tight wrist strap and perform keying tasks. A wrist restraint may also restrict blood flow to the hands. And, if worn firmly over a long period, the immobilization it causes may also produce some loss of hand strength. If the workplace is properly redesigned, sharp bending of the wrist when using the keyboard can be eliminated. This redesigning involves adjusting the height of your seat to suit both your height and the height of the desk. In this way, the keyboard can be brought to or just above elbow level—a comfortable level.

Wrist restraint devices cannot compensate for a poorly designed workstation. Still, many employers advise their workers to wear them because they are quick and easy to implement, they require no equipment or workstation modification, they do not disrupt operations as workstation redesign does, and they shift the blame for CTS problems from the employer to the worker. A wrist restraint is an addition to the structures of your body. Many employers may also honestly believe that wrist restraints are the solutions to CTS problems. This is a false belief. It is fostered by the lack of education about the relationship between work and strain in the body and by the powerful impact of advertisements for the sale of these devices.

BURSITIS

In the shoulders, elbows, and wrists, the rubbing motion and pressure of a tendon on a bone or ligament may be cushioned by a flattish sac called a **bursa.** This sac is lined with synovial membrane and filled with synovial fluid. One frequently injured bursa can be found at the shoulder joint between the tendons of the **rotator cuff** muscles and the head of the humerus (the upper-arm bone). Baseball pitchers are at a high risk for this disorder. Sustained tension or excessive rubbing friction of tendons over a bursa may subject the tendons to so much wear that they may become roughened and irritate the adjacent bursa. Eventually, the bursa may become inflamed—a condition called bursitis.

Bursitis is not common among computer workers, but a possible cause of the cases that do occur in computer work is prolonged lifting of the shoulders, as when working with the arms extended or raised and suspended in midair. This happens when the keyboard is either too far away, too high, or both. Very short people are more likely to experience this problem. When your arms are reaching forward or upward, their weight (10% of your total body weight) must be supported by contractions from shoulder muscles. The tensed tendons of these muscles then exert powerful pressures on the shoulder bursa. If you are forced to lift your arms to use your keyboard, you can reduce shoulder muscle contraction by resting the forearm on the edge of the desk. But this edge must be rounded, padded, or fitted with a special forearm rest. Otherwise, it may create too much pressure on the tendons, nerves, and blood vessels under the skin of the wrist in contact with the desk.

EPICONDYLITIS

Tennis players suffer from inflamed tissues in their elbows, a condition believed to be caused by the constant action of hitting the tennis ball with the racquet. Each hit requires strenuous contractions of the muscles that rotate the forearm. The backhand stroke is considered worse than the forearm stroke in this respect. The strain accumulates over months or years. Eventually, inflammation may result at the elbow end of the tendon or muscle where it joins the projection (the **epicondyle**) of the upper-arm bone (the humerus). This disease was named epicondylitis after the exact part of the elbow that was affected. Its more common name is tennis elbow, after the game that was thought to produce it. The disease is also common among workers who use certain hand tools, such as manual screwdrivers, which force them to rotate the forearm constantly. Some people who work at computers complain of similar pains in the elbow.

When using a keyboard, your forearm must be turned so that your palm faces the keyboard. The constant contraction of the **pronator muscles** (those that turn the forearm), coupled with the incessant contractions of the muscles that power the fingers, may be enough to cause inflammation at the elbows in some people. Also, bending your wrist will increase the strain in the tendons and may exacerbate the debilitating conditions that lead to tennis elbow.

POINTS TO REMEMBER

- Repetitive motion injuries (RMIs) are cumulative trauma disorders caused by excessive motions of a body part. The combination of unnatural bending of joints, sustained muscular contractions, repetitive finger motions, and inadequate rest is likely to make your body more susceptible to RMIs.
- Tendinitis, tenosynovitis, and carpal tunnel syndrome are common types of RMIs in the hand and arm. Epicondylitis of the elbow and bursitis of the shoulder are less common cumulative trauma disorders.
- Each time one of your finger taps on a key, its muscle(s) contracts and its tendon(s) tenses.
- CTS is accompanied by intense pain in the wrist, especially when it is bent. Swelling, numbness, cramps, and partial loss of hand strength and control are also common symptoms.
- Wrist splints do not prevent or delay the onset of CTS caused by repetitive motions if you continue to perform the same work.
- CTS can also be caused by other factors not related to work, like use of oral contraceptives, menopause, previous fractures of the wrist, diabetes, and rheumatoid arthritis.

CHAPTER 13

THE KEYBOARD

The universal acceptance of a product or tool does not necessarily mean that it is free of faults. Take the computer keyboard, for example. Few people realize that its peculiar design imposes considerable strain in your hands and arms—strains that can lead to tenosynovitis, tendinitis, or carpal tunnel syndrome.

The common keyboard is so widely accepted that many people find it hard to blame its design for cumulative disorders of the hand. However, the physical dimensions of the keyboard, its shape, and the arrangement of its keys do in fact cause problems for people who work at computers. While keying, your arms and hands are forced into constrained and stressful configurations that are not conducive to prolonged work. Your health eventually suffers as wear and tear of the musculoskeletal structures occurs gradually.

Keyboard Design Versus Hand Design

The geometry of the keyboard and the anatomy of the hand are incompatible in some important respects. The keyboard is rectangular in shape with four horizontal rows of keys for numbers and letters. The space bar is in a fifth row. The rows descend slightly from top to bottom, and each row lies in a perfectly straight line. Each finger must be placed in a specific position to operate the keys properly. This design forces the arms and hands away from their natural positions of minimal stress into awkward, stressful positions.

To better understand this, you can perform the following experiment. Stand and allow your arms to hang freely at your sides. Now bring your lower arms upward until they are horizontal, keeping the upper arms alongside your trunk. Your upper arms and forearms should now be in the typical typing position, but not your hands. The forearms should be parallel to each other with the palms partly facing each other, and your fingertips should form a curve, with the little finger well below the thumb and index finger. In this position your hands and fingers cannot use a keyboard, but the tension in the finger muscles and tendons is minimal. Any rotation of the forearm or bending of the wrist will, therefore, increase this tension. Unfortunately, this is exactly what you must do to use the keyboard.

In order to use a keyboard properly, you must make three major adjustments to this minimal stress position. Your hands must be moved toward each other to position the fingers over the alphabetic keys. They must be reoriented so that your palms face the keyboard and fingers point downward towards the keys, and your wrists must be bent sideways towards the little-finger side to bring each finger over a specific key. These motions are achieved by contractions of the corresponding muscles in your forearms and hands. The resulting increased tension in the muscles and their tendons must be maintained for as long as you are keying at the computer. This is what builds up muscular strain.

Keyboard Width and Shoulder Width

The keyboard width and the width of your shoulders are not the same (see Figure 16). You can get a good idea of the great disparity between the width of the alphabetic keyboard and people's shoulder widths by considering the following information. The alphabetic characters and crucial function keys lie within a rectangle eleven inches wide. When you use the keyboard, the fingers that are furthest apart are the two little fingers; the left one is over the "A" key, and the right one is over the ":" key. These keys are only seven inches apart. If the width of your shoulders was seven inches, then your fingers would fall over the alphabetic keys when the two forearms are parallel and relaxed, as shown in Figure 16. But your shoulders are much wider than that. The shoulder width of U.S. women averages about sixteen and a half inches, and the smallest 5% of them are below fifteen inches. Men are, of course, in a worse situation. Their shoulder width averages nineteen inches, with the smallest 5% of them below seventeen and a half inches. These figures mean that a man of average shoulder width must move each hand inward by about six inches, and a woman, about five inches.

Figure 16.
Top view of an operator in a typical typing posture, indicating the mismatch between the keyboard and the operator's sizes.

Inward movement of the hands to achieve the typical typing position requires contractions of shoulder muscles. The subscapularis muscles of the shoulders must bear the burden of this hand-keyboard mismatch. The contraction level is not that great, but it must be maintained throughout each day of your working life. The cumulative strain is, therefore, significant.

Positioning Your Fingers

In normal typing methods, the tips of your fingers (except your thumb) must be positioned just above the "A", "S", "D", "F", "J", "K", "L", and ":" keys. Since these keys are at the same horizontal level, the tips of all eight fingers must be brought to the same level. From the relaxed arm position, this is achieved by rotating the forearm so that the thumb side of the hand is lowered and the little-finger side is raised. The rotation is considerable when using a keyboard—about 90 degrees. Two forearm muscles that cross the elbow, the **pronator teres** and the **pronator quad-**

ratus, must contract to rotate the forearm. They remain in a state of tension for as long as keying is performed. For many people this means six to eight hours a day. In addition, the rotation creates great mechanical stress at the elbow where the ends of the two forearm bones meet. This joint is rotated almost to its limit. Some people try to lessen the strain from forearm rotation by raising the upper arm sideways, but this merely transfers the strain to the shoulder and places the hand in an awkward position.

Even with the two hands in close proximity and the fingertips at the same horizontal level, the hands are still not in their proper typing position. The fingertips are not yet over the appropriate keys. Instead, they are approximately diagonal (slightly curved) across the keyboard. One more adjustment must be made. The wrist must be bent sideways towards the outside. The muscles that do this are the **extensor carpi radialis longus** and the **flexor carpi radialis.** Both run from the elbow joint to the metacarpal bones in the palm of the hand. Bending the wrist also makes the carpal tunnel smaller and increases the pressure on the tendons and median nerve.

The tendons of the muscles that position the hands over the keyboard originate at the elbow. Working at the keyboard over a long period of time will certainly build up strain at this joint. The sustained contraction of these muscles is felt as tightness and pain at the elbow by the end of the workday.

Another detrimental consequence of the traditional keyboard design concerns the distance between the "A" and the "Shift" keys for the left hand. In the standard typing position, the little finger of the left hand is positioned over the "A" key, and the right hand little finger over the ":" key. In order to make upper case letters you must depress a "Shift" key. This is normally done by moving the little finger outward and rearward without moving the rest of the hand. This action, as simple as it may seem, can easily fatigue muscles and tendons when done often. An imbalance in the workloads of the two little fingers is also created because the little finger of the right hand must move a little further than the little finger of the left hand. Look at your keyboard; you will notice that the distance between the ":" key and the "Shift" key for the right hand is greater than the distance between the "A" key and the "Shift" key for the left hand.

Finger Force for Keying

The modern electronic keyboard has undergone many changes. One of these is the reduction in the finger force required to depress a key on the electronic keyboard. Some people may think that this should help reduce overall strain in the hands, but it does not. Incidences of cumulative trauma disorders in the hands have increased drastically with the use of the computer's electronic keyboard. This paradox suggests that there are other, more important factors at work. As I have already mentioned, maintaining the arms and hands in constrained positions for prolonged periods and rapid keying are two of the worst factors. Another factor is the way the keys must be depressed. When a key is depressed there should be some cue to tell you when to stop applying force. Otherwise, the finger will press the key until it hits the base. Some experts have suggested that this sudden obstruction to the movement of the key sends a miniature shock wave back along the fingers and into the wrists. While a few of these shock waves may be harmless, the accumulation of the vibratory force from several thousand each day is likely to affect the tendons and nerves adversely.

Modern keyboard designs incorporate two cues to solve this problem—a tactile one and an auditory one. In the first case, when you depress a key you feel a change in the smoothness of its motion like a miniature bump about halfway down its travel. In the second case, the change is accompanied by a clicking sound. These two cues warn you that the key has been activated and that you should stop applying force. After a time, your reaction to the cues becomes automatic. By this method, the sudden impact of the key at the end of its travel and the miniature shock wave it generates can be avoided. But vendors still sell cueless keyboards, and many people still use them, perhaps because they are inexpensive.

Arrangement of the Keys

The arrangement of the keys on the keyboard is another factor that can contribute to excessive strain in the tendons of the fingers. The alphabetic keys in the earliest keyboards were arranged in three main ways—circular, linear, or rectangular. But the rectangular arrangement gained more widespread acceptance. For decades now, the speed at which a typist uses the keyboard has been one of the two main criteria that determined his skill as a typist. The other is, of course, accuracy. But the peculiar arrangement of letters on a typewriter seem to inhibit both of these. The first five letters in the top row of keys are "Q", "W", "E", "R", "T", and "Y". This is why your keyboard is called a "QWERTY" keyboard. Actually, this QWERTY arrangement was specifically designed to reduce the speed of typing. This paradox resulted from a conflict between the speed capabilities of the earliest typewriters and the typing-speed capabilities of typists. Typists were able to type faster than the speed at which the keys of these older machines worked. The letters used to be arranged in alphabetical order, but this caused problems. In these typewriters, each key was attached to a lever arm or type bar that acted like a hammer when depressed. The lever catapulted onto the paper when finger force was applied to the key and was retracted when the finger force was removed. However, if two adjacent keys were depressed in quick succession, the two levers often became entangled with one another. This problem was solved by separating the keys that were frequently used in quick succession. The result was the ubiquitous QWERTY arrangement, designed by Charles Latham Sholes in 1873.

The workload is not equally divided between the two hands when you use the QWERTY arrangement. The left hand keys are used about 57% of the time and the right hand ones 43%. In addition, in right-handed people (who constitute about 93% of the population) the left hand is about 7% to 10% weaker than the right hand. In left-handers the two hands are of about equal strength. So for most people the left hand is subjected to greater mechanical stress using a keyboard. Most of the hand problems among computer typists have occurred in their left hands. If the work-load distribution on the hands were reversed, it is likely that hand fatigue would occur less frequently than it does now.

There are other problems with the standard QWERTY keyboard. The fingers must jump back and forth excessively from row to row. They travel twelve to twenty miles per day by these movements. The work assigned to the different fingers is also not compatible with their skill and strength.

A Better Keyboard Design?

The above description of the standard QWERTY keyboard implies that the key arrangement and general keyboard design are crude. The decision to retain the QWERTY keyboard despite its flaws has been motivated largely by economics—costs for new keyboards, cost of errors while learning to use them, and administrative costs for the changeover. Tens of millions of people would be required to learn a new style of typing with a new, although better, design. This would be no easy task. The finger movements of a competent typist are automatic. The actions are like conditioned reflexes. Replacing them could cause errors early in the changeover, which could be intolerable in today's workplace. The time it would take you to reach your former level of proficiency with the new keyboard may be too long for the liking of today's managers. They tend to demand immediate results and excellent performance at all times.

The costs for the new keyboards could be recouped in the long run through enhanced productivity and decreased musculoskeletal disorders of the hands, arms, and upper body. But it is no easy task to convince management to make the change. In our society, promotion opportunities and job security for professionals are based on short-term performance, and decisions that can be profitable in the long term are often sacrificed for those that have visible short-term gains. Then there is also the normal resistance to change. It is difficult to get support for changing the way our world has been doing things successfully since 1873.

The best keyboard would be one that you could use with your arms in a more natural position. This position is the one achieved in the experiment that was described earlier in this chapter.

The good news is that such a keyboard has been developed. The basic problem was solved as long ago as 1926 by a German researcher named Klockenberg. He made a number of design suggestions for reducing strain in the arms. Splitting the keyboard into two parts and separating them so that the hand and wrist can

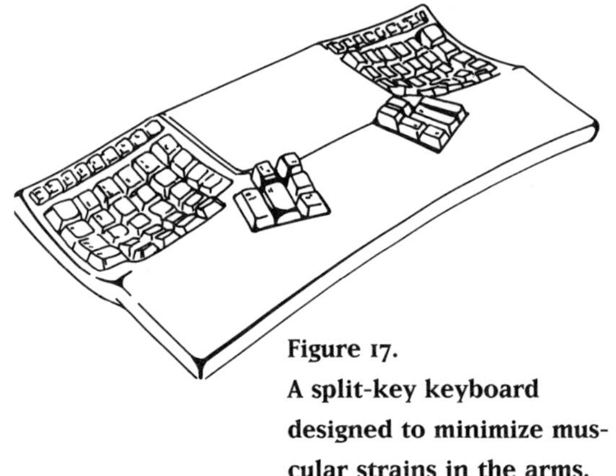

Figure 17.
A split-key keyboard designed to minimize muscular strains in the arms.

stay in a more natural position was the most revolutionary change. A few alternative keyboard designs, using the Klockenberg concept, have been tested by several researchers. The results have been encouraging. Strains in the hands have been reduced significantly, and users have welcomed these new keyboards. One of these, called the Maltron keyboard, has become available commercially fairly recently (see Figure 17).

In the Maltron keyboard, the keys are arranged in two separate clusters. The right hand uses the keys on one cluster and the left hand uses the other cluster. The clusters are separated by a small distance so that your hands need not be brought toward one another. The keys in the two clusters are not at the same horizontal level either. They are curved or dished to conform to the natural contours of your fingertips and are angled downward and away from the user. So there is no need to rotate the forearms or turn your wrists sideways significantly to bring the fingertips into a horizontal line. The tension generated in the muscles and tendons of the arm when using a traditional keyboard is then eliminated. You should feel less fatigued in the arms, neck, and shoulders at the end of a workday using this keyboard instead of the traditional one. However, one source of muscular strain will still be present. This is the strain resulting from rapid and incessant movement of the fingers. This brings us to another point.

With an improved keyboard, you should be able to work faster and for longer periods, which could lead your employer to increase your workload. This is exactly what happened when the computer replaced the typewriter. So the relief gained from the elimination of constrained arm and hand postures may be supplanted by the strain from greater speed and volume of work.

POINTS TO REMEMBER

- The design of the keyboard is partly responsible for repetitive motion injuries.
- The keyboard's dimensions and shape do not conform to the dimensions of your shoulders and the shape of your hands.
- The small width of the alphabetic keyboard forces you to move your forearms and hands inward toward one another, resulting in muscular strain at the shoulders.
- The flat arrangement of each row of keys forces you to rotate your forearms inward, resulting in strain in the pronator muscles of the forearms and in the elbow joints.
- The linear arrangement of each row of keys forces you to turn your wrists outward, resulting in strain in the finger tendons.
- The well-known QWERTY keyboard was developed in 1873 to slow down the speed of typists and prevent jamming the mechanical arms of the typewriter.
- A new keyboard has been designed to eliminate these problems. It is split into two parts so that each hand gets its own keyboard, and each part is angled and tilted to prevent wrist deviation. Each part is also dished at the top to prevent forearm rotation.

CHAPTER 14

LIGHTING, VISION, AND EYESTRAIN

We depend on our eyes for about 90% of our activities during waking life. Visual work in modern societies has become intensive, and nowhere is it causing more widespread problems than in the computer workplace. In the U.S., more than sixty-five million people work at computer terminals every day, and many of them suffer from vision problems caused by their jobs. A recent survey by the American Optometric Association indicates that over eight million people who visit their optometrists suffer from job-related problems. Complaints such as irritated and watery eyes, burning or soreness, inability to focus properly (blurred and double images), seeing colored fringes on objects, headaches, and aches in the eyes are all common among computer users. One study, published by the prestigious *Human Factors* journal in 1981, found that eye complaints from clerical workers using computers increased by 26% to 36% compared to those who did not use computers. Changes in color perception increased by 31%, eye irritation by 27%, burning in the eyes by 26%, blurred vision by 36%, and general eyestrain by 31%. In the worst cases, some of these symptoms even persisted at home after work. In addition to problems with the eyes, you can also suffer from aches and pains in the neck and shoulder area and even in the lower back from trying to compensate for visual difficulties.

We can learn how to prevent these problems by examining the various conditions that are associated with or cause them:

- Conditions causing difficulty in focusing your eyes.
- Conditions that strain the eye muscles and eventually lead to headaches.
- Conditions that force your head and neck into constrained postures.

Researchers and eye specialists have indicated that computer work causes no permanent damage to the eyes and that the symptoms of eye problems are transient and reversible, often after a good night's rest. However, even though there may be no permanent damage to the eyes and the associated nervous system, you should still be concerned about visual problems from computer work. No one wants to work in misery, even if it disappears after a night's rest. These maladies recur regularly, and medical pronouncements that they are not permanent offer little consolation to the many affected operators. Management should be concerned as well because visual discomfort translates into lower productivity. Visual strain affects your ability to concentrate on written material on documents or the computer screen and understand it.

Musculoskeletal Problems and Visual Defects

While it may be obvious to us that problems with seeing affect the eyes and are associated with the quality of lighting, it is less obvious that some of the aches and pains in other parts of the body are the result of visual problems in the workplace. In fact, studies have shown that many people who complain of pains in the neck and back also complain of vision problems at the same time. Glare, poor lighting conditions, and wearing corrective glasses or contact lenses at the computer workstation may cause you to adopt constrained or unaccustomed postures in order to read properly from the screen, keyboard, or documents.

Eventually you suffer from aches and pains that may not seem to be related to visual problems. This is why it is important to know how these problems occur.

We have seen in previous chapters how, in order for your hands to use the keyboard properly, the rest of your body must be properly positioned. We have also seen how tension and fatigue in muscles and compression and other phenomena in joints are magnified when the body is forced to adopt constrained postures to get the hand to work properly. In this chapter and the next, we will look at vision and try to understand how its defects and poor lighting conditions cause or contribute to these problems.

The Structure of Your Eyes

This simple description of your visual system serves only to illustrate the basic concepts of how you see things so that you can understand the difficulties of working at computer workstations.

Two parts of your body are responsible for seeing, your eyes and your brain. Your eyes form images of objects from outside the body, and your brain interprets these images. An image is formed when light is reflected from an object then enters the eye and becomes focused on the eye's inner rear wall (the retina). The brain then interprets the image so that you know what you're looking at.

Your eyeball is a slightly flattened sphere, barely smaller than a Ping-Pong ball. It is held in the bony eye socket by a set of six extrinsic muscles that run from its outer surface to the wall of the orbit (see Figure 18). The contractions of these muscles rotate the eyes within the orbit and allow you to change your line of vision without moving your head. They allow you to see within a horizontal range of 208 degrees and a vertical range of 130 degrees when both the eyes and head are stationary. The wall of the eyeball is composed of three coats of cells, the **sclera** and **cornea** on the outside, the **choroid** in the middle, and the **retina** on the inner rear wall.

The sclera is white and opaque and covers most of the outer surface of your eye. The cornea is clear and translucent and covers only one-sixth of the outer surface, at the front of the eye. It is the window of the eye. The choroid, which is under the sclera, continues to the front of the eye and terminates in an enlarged muscular section called the ciliary body or cilary muscle. This controls focusing of the eye lens.

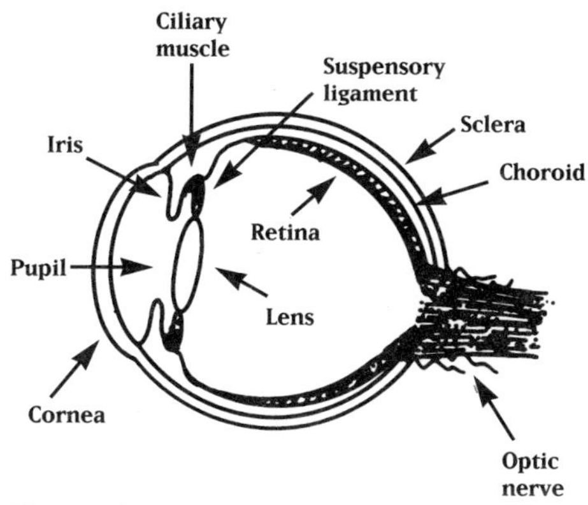

Figure 18.
A section through the eye, showing its most important parts.

Just behind the cornea and in front of the lens of the eye is a circular wall of muscles called the **iris** that contains a circular opening, the **pupil**, in the center through which light enters the eye. The iris consists of two sets of muscles—one running concentrically (in circles) and the other radially (from the center toward the outer edge). Their contractions change the size of the pupil and regulate the amount of light entering the eye. The pupil is dark because when light enters the eye it is not reflected back out.

The **lens** of the eye is located behind the iris. It is essentially a layer of translucent cells curved outward on both its front and rear surfaces. By changing the sharpness of this curvature, the eye can focus on objects at different distances. The lens is held in position by ligaments, called suspensory ligaments, that connect at their other end to the muscular ciliary body. The contractions of the ciliary muscles influence the pull of the suspensory ligaments on the lens, which affects the shape of the lens.

The retina lines the inner wall of the eyeball and is continuous with the **optic nerve**, which connects the eye to the brain. The optic nerve itself runs into other nerves in the brain. The retina forms the image of any object you look at. When light hits the retina, it generates bioelectric impulses that are relayed to the brain via the optic nerve in the same way that TV cables relay pictures from the cameras to your TV at home. The brain then interprets these signals and sees the object in the environment.

The inside of the eye is filled with a clear watery fluid in front of the lens, and a clear jelly-like substance behind the lens. These two substances inflate the eye and maintain its spherical shape.

Four basic visual processes occur when you look at an object, such as a computer screen:

- The eyeball is rotated by its extrinsic muscles to position the eye so that light reflected from objects can enter it.
- The pupil size is regulated to allow the appropriate quantity of light to enter the eye.
- The focusing power of the lens is regulated by a change of its thickness so that light can be focused on the retina.
- Bioelectric impulses are generated by the image on the retina and travel to the brain via the optic nerve.

If any of these processes fail to work properly, vision deteriorates and visual strain results. Physiological vision defects, as well as certain work and environmental conditions, can cause one or more of these processes to work improperly. The remaining sections of this chapter will describe how adverse conditions in the computer workplace stress the visual system and cause one or more of these processes to fail to work properly and how this failure can lead to eye fatigue and even musculoskeletal aches and pains.

Focus of Your Eye Lens

As light travels from outside your eye to the retina, it is refracted twice: first by the cornea, then by the lens. Most of the bending or focusing (about 70%) is done by the cornea. But the degree of bending by the cornea is fixed and does not necessarily focus light on the retina. The lens completes the process by fine tuning the focusing because it can vary its power of focusing.

You will notice that when you look at distant objects, things close to you are blurred, and when you look at close objects, distant objects are blurred. This is because the eye can only focus on things within a certain range at any one instant. This range is called the **depth of focus.** In order to focus on things out of that range, the eye lens must change its focus, a process called **accommodation.**

The eye lens changes focus by changing its shape. Recall that the lens is suspended in position by ligaments connected to its periphery and that the ligaments are connected to the ciliary muscles. The contractions and relaxations of the ciliary muscles change the tension in the suspensory ligaments, which then change the shape of the lens. When the ciliary muscles contract, the suspensory ligaments relax and vice versa.

Light from objects closer to the eye must be more sharply bent than light from more distant objects. When looking at distant objects, the ciliary muscles relax while the tension in the suspensory ligaments increases and pulls on the periphery of the lens. The lens is then flattened and its focusing power weakened enough to focus the light on the retina. When looking at objects close up, the ciliary muscles contract while the tension in the suspensory ligaments is released. The lens then bulges in the middle and its focusing power increases. The light from close objects is then focused on the retina.

Computer work is essentially close-up work. The eyes maintain focus on the screen, keyboard, or documents for several hours each day. This means that the ciliary muscles, which help to control lens focusing, are under sustained contraction. As you grow older, these muscles become weaker and you lose some close-focusing ability. You must then read from a farther distance. Screens and documents that are close enough for younger people are likely to strain the ciliary muscles of older eyes.

Visual accommodation at the computer workplace can be troublesome if it requires frequent and quick changes of view among three visual displays—the screen, keyboard, and documents. In the typical computer workplace, these three displays are often positioned at different distances from the eye because of insufficient space on the desk. Therefore, every time you look from one to the other your eyes must change focus. In youth, this is hardly a problem. But, as you get older, you experience greater difficulty in changing focus. Beyond about forty years of age, your eye lens gradually becomes less flexible and loses some of its ability to change shape. The lens then takes longer to change focus, and you tend to perform more slowly at computer jobs that require quick and frequent changes of focus. However, your ability to perform at jobs requiring significant mental activity and decision making is not affected by accommodation problems.

Frequent head movements at the computer workstation are a consequence of the placement of the three main visual displays—the screen, the document, and the keyboard. The movements may be reduced or eliminated if the displays are placed as closely as possible, and frequent accommodation may be reduced if the screen and documents are placed at the same distance from the eyes. There is wide variation in the distance at which different people set their screens from their eyes. Studies have discovered distances ranging from 24 to 36 inches. It is difficult to recommend any specific value. You should experiment by trial and error to determine the best distance from your eyes to place your screen, documents, and keyboard. It will probably be close to 24 to 36 inches. But if it falls outside this range, don't worry. Your vision may simply be different from most of the population. Your optometrist can give you solid advice here.

Work Environment

Problems with vision in the computer workplace lie mostly in the nature of the screen. Before the advent of the desktop computer, typists viewed their work on paper. The print was black, or nearly black, and the contrast against the white paper was excellent for reading. In fact, this black print on white paper was the universal display for almost all printed material—books, newspapers, magazines, and so on. So the eyes were adapted to this kind of medium. In addition, the legibility of typed documents was enhanced when typists used bonded paper, which reflects little light. However, with the advent of the computer, the mirror-like screen replaced typewriter paper, and the nature of visual work in the office was changed forever.

The screen reflects light from various sources in the room, and your eyes are almost always in the path of the reflected light. Brief spells of work under difficult visual conditions would do little harm, but prolonged and intensive reading will inevitably lead to eyestrain.

Lighting

To prevent eyestrain in computer work, room lighting must be a high quality. But this does not necessarily mean extremely bright lighting. Poor quality of workplace lighting and poor positioning of the computer screen, documents, and keyboard are all major causes of glare, high contrasts, and reflections. Workplace lighting that is too bright, too dim, or that is reflected can easily lead to eyestrain.

Excessively bright light entering your eyes does not produce a sharp, clear image on the retina. The effect is similar to overexposure of film in a camera. But the amount of light entering your eyes is automatically limited by the contraction of the muscles of the iris. The concentric muscles contract while the radial muscles relax, reducing the size of the pupil. Prolonged work in an excessively bright workplace will lead to strain in the concentric muscles. Eventually, you may be left with aches in your eyes or head.

Dim or insufficient lighting can cause strain in the radial muscles in the iris and lead to aches in the eyes and head. In dim lighting, the radial muscles of the eye contract while the concentric muscles relax to dilate the pupil so that more light can enter the eye. As long as you continue to work in dim light, these radial muscles remain in a tensed state.

People tend to compensate for dim lighting by bringing their eyes closer to the display. But this means bending your neck and thrusting your head forward. As I explained earlier, this increases the mechanical strain in the muscles and spinal discs in your neck. Sometimes, ambient lighting may be so dim or the writing on the document so weak that you are forced to bring your head closer to the document by bending both your trunk and neck forward. This increased mechanical strain spreads to the lower back.

Glare

Glare refers to excessive amounts of light entering the eye and impinging on the retina. As a result, a clear image cannot be formed on the retina. The retina becomes overexposed. Glare is the most common disturbance in vision in the computer workplace and the most difficult to eliminate. Ask any typist who has changed from the typewriter to the desktop computer about new problems in reading from the screen, and he will not fail to mention glare. Glare is directly responsible for eye fatigue and headaches and indirectly responsible for musculoskeletal strain.

There are several sources of glare in computerized offices. These include bright light from the sun entering through windows; task lamps close to your eyes; intensely illuminated characters on the screen; reflections from smooth, polished, or light-colored surfaces such as desk tops; white paper, walls, and polished floors; and, worst of all, reflections off the computer screen. Glare that is caused by light entering the eyes directly from a source, as when you face a bright window or task lamp, is called **direct glare**. Glare that is caused by reflections is called **indirect** or **reflection glare**.

Problems with reflected glare are almost ubiquitous in the modern office. The computer screen and polished or smooth desk tops are among the strongest reflectors of light. But the surfaces of certain types of keyboard keys and smooth white paper are also appreciably reflective. In one study of fifteen open-plan offices with a total of 519 employees, 23% said that they were affected by glare. Of these, 36% thought that windows were the fault, 25% blamed lamps, 18% said it was polished tabletops, and 17% pointed to excessively bright lights. These figures are typical of many offices.

Visual Field and Line of Sight

Unwanted light or glare from any source can enter your eyes if your **visual field** lies in its path. The visual field is the area in front of your eyes that you can see when your head is kept motionless. It covers a cone of angle 40 to 70 degrees about the eye. This is a fairly large area, and it makes your eyes susceptible to glare from a wide area of the room. While looking at the screen ahead of you, for example, bright lights from ceiling lamps, task lamps, or windows can easily fall into your visual field. It is difficult to eliminate this type of glare simply by reorienting your screen and changing your seating arrangement. The sources—lamps or window light—must be kept out of your visual field.

There is a tendency in modern offices to emphasize aesthetics, which often involves the use of highly polished furniture and other smooth surfaces. While they may be soothing to the mind, these surfaces are a scourge to the eyes of those who perform intense visual work because they reflect too much light.

Ideally, we would like to have a workplace where glare is completely absent. But the science of designing environments to conform to human visual capacities lags behind our knowledge of the science of illuminating workplaces. We are still unable to design workplaces to eliminate glare, and we seem to be far from achieving this. Sometimes a new and better design cannot be implemented because it is too costly. For now, we must be content with controlling and minimizing glare.

Combating Glare

The most troublesome reflections are from the computer screen. The screen acts like a mirror and reflects light from a variety of sources. These reflections are the background images that you see in the screen. They wash the screen with light and reduce the contrast between the illuminated characters and their background. Since a certain level of contrast is necessary for easy discrimination of characters from their background, reading washed characters from the screen can be difficult. Your eyes and brain are forced to work harder to compensate for the loss in contrast. This may eventually lead to fatigue in the optic nerve and in the eye muscles. But, if your eyes are not in the path of reflected light, your vision will not be degraded.

Almost all methods for combating glare are based on three principles—reducing the intensity of the light from their sources, changing the direction of the light, or changing the direction of the reflected light. Sources of glare that are controllable economically and that should be controlled include:

- Overhead light
- Task light
- Window light
- Screen reflections
- Reflections off surfaces

One reason there are so many problems with reflection glare is that most offices were not designed for computer work. Instead they were designed for traditional office work with typewriters. Computers were simply moved into these offices to replace typewriters, often without any change in the physical environment. Traditional office lighting was adequate for traditional office work with typewriters, but it causes severe problems for work with computers. Overhead light in the computer workplace can produce either direct or reflection glare, and when it is bright it can produce contrasts, reflections, or shadows that are too deep for seeing things easily.

Direct glare can occur in an office if the ceiling is low enough for light from the bulbs to fall within your visual field, especially if the bulbs are not placed in recesses in the ceiling. If you spend more time looking at the screen than at documents, you will be more susceptible to such glare. Reflections of overhead lights by the computer screen are the most common cause of indirect glare. It is perhaps the most difficult type of glare to eliminate. Images of these lights in the screen always seem to be staring at you, no matter how you orient the screen or your head.

The amount of light from overhead lamps on a particular surface is called the illumination or illuminance of the surface and is measured in units of **lux**. Illumination at desks and other work surfaces in most computer offices is 400 to 700 lux. By comparison, bright sunlight is about 150,000 lux, and light next to you from good street lighting at night is about 10 lux. You can measure the brightness at your desk with a light meter, which can be bought from manufacturers of environmental measurement instruments.

How bright should your computer workstation be? It is unrealistic to recommend a specific level because it depends on a number of factors. The most important are individual preferences, type of job, and seating arrangement with respect to the source of illumination. Individual preferences are due to different visual acuity levels and reading habits. The type of job determines which display (screen, document, or keyboard) you will look at most of the time while working, which determines the level of illumination. Guidelines by experts vary widely, but most fall between 300 and 500 lux. The Human Factors Society/ANSI guidelines recommend 200 to 500 lux, with task lighting directed at paper copy if necessary.

Research suggests that operators who use higher levels of illumination have fewer problems with visual accommodation changes. One writer has suggested that this is probably because, in higher levels of ambient lighting, operators can see more as they look around the room more, at a distance from the computer workstation. This encourages looking at distant objects, whereas dim levels discourage looking around. This distance viewing relaxes the ciliary muscles that control the shape of the eye lens, keeping them healthy.

Common methods for reducing glare from overhead lighting include repositioning the lamps, dimming overhead lighting, and using task lamps in combination with dimmed overhead lighting. Repositioning overhead fixtures and lamps is seldom practical since it may involve elaborate and expensive rewiring. In addition, you must be sure that the new positions will reduce glare significantly, and that is no easy task. Positioning of overhead lamps should be determined at the design stage of a building, not after the computer workplace has already been set up.

Dimmer switches make dimming overhead lighting easy and convenient. Dimming does not create problems with reading on the screen because you can adjust the brightness of the screen characters to achieve a suitable level of contrast against the screen background. This can easily be done by turning the brightness control knob located at the side or at the bottom of the monitor. All modern desktop computer monitors have this feature. However, dim lighting makes it difficult to read from documents and the keyboard. This is where you can use a task lamp.

A task lamp is usually bright, and care must be taken in positioning it. To avoid direct glare, do not bring it too close to your face or within your field of view. The best position for the task lamp is at your side, a little above your head. If it is positioned in front of you, it can cause glare by reaching your eyes directly or by reflecting off shiny surfaces in front of you. If it is positioned behind you, it can reach your eyes by reflecting off the screen. You should also use a lamp shade that is large enough to shield the bulb from your eyes to eliminate direct glare.

A task lamp placed too close to the face can also create a

bright area at your workplace and make surrounding areas seem dark by contrast. This can cause eye muscle fatigue when you look from one of these areas to the other alternately. In general, the area at the center of your visual field should not be more than ten times as bright as the area immediately surrounding it. You must also consider the effects of task lamps on coworkers. If your workstation is isolated by walls, your task lamp will not bother others in the office. But in an open-plan office, bright light from your task lamp can easily fall within the field of vision of those nearby.

Different Types of Work and Lighting Needs

Conflicts about the use of dim overhead light can occur among different types of computer users in the same room. In different types of jobs, people spend different amounts of time reading from documents, the screen, or the keyboard. Reading from the screen is affected less by dim overhead lighting than reading from documents or the keyboard. CAD (computer-aided design) or conversational operators, for example, spend much less time looking at documents and more time looking at the screen than data entry or word processing operators so they are less affected by dim overhead lighting.

The problem with different lighting needs for different types of computer jobs can be solved by providing different offices for different types of operations. But this may require greater overall office space, which, in turn, means greater operating costs for the employer. This method may be too costly for a small company with a small number of operators who perform different kinds of work.

Because of possible glare and contrast problems associated with a task lamp combined with dim overhead lighting, normal levels of overhead illumination (about 300 to 500 lux) may be better. Some people prefer sunlight, not for its brightness but to ward off feelings of being confined to a dungeon. Unfortunately, few offices were designed to make significant use of sunlight. However, you can make use of sunlight in your home office—if you can eliminate its troublesome glare.

Window Light

Glare from window light can be avoided by positioning your desk so that the window is to your side when you sit. Avoid facing the window or having your back to it. Facing a window or an unshaded glass door brings sunlight directly into your eyes, while keeping your back to it may cause its reflection from the screen. The larger the window or the lower its position, the easier it is for this kind of glare to affect you.

The use of window blinds made of fabric, aluminum sheets, or plastic sheets is now popular. But the reflections of some of the plastic and aluminum blinds are easily seen in the screen. These blinds reflect enough light to produce secondary reflections from the screen. This is because they are smooth and highly polished. Ironically, the most popular blinds are of the more highly reflective colors—white, cream, and alabaster. These colors are often chosen for their match with walls and floors and not necessarily for their ability to prevent reflection of light. Fabrics absorb more light than smooth metal or plastic sheets and are better for window blinds.

Document Holder Orientation

The orientation of a document holder depends on the quality of lighting. Light can easily be reflected off documents, especially if the paper is smooth and white. Bonded paper reflects much less light. Glare can be alleviated by reorienting the document holder so that the reflected light is turned in another direction. The document holder should be rotated vertically if the bothersome light source is above the holder or sideways if it is at the side of the holder. Almost any document holder can be rotated sideways, but not all can be rotated in the vertical plane. You should use one that can be rotated both ways. Without vertical rotation, you may be forced to bend your head and neck to get your eyes out of the path of reflected light. Sometimes, someone forgets that she can rotate the document holder and rotates her neck sideways, bends it up or down, or leans her body sideways to avoid reflections from documents. In these postures, excessive strain will be imparted on the lumbar spine or neck.

Controlling Screen Reflections

Reorienting the computer screen is a common method of combating reflections from overhead lights and windows. Remember when you were in elementary school and experimented with mirrors and sunlight. You probably discovered that you can rotate a reflected beam of sunlight by simply rotating a mirror in your hand. The same principle is employed in rotating the computer screen to avoid reflections. The light that is reflected from the screen into your eyes can easily be turned in another direction, away from your eyes, by rotating the screen. Rotating the screen horizontally and vertically is facilitated by the swivel-tilt base that modern monitors sit on.

As you rotate the screen you will notice the undesirable reflection moving to a corner and disappearing off the screen. However, you may not be able to totally eliminate screen reflection glare in this way. The reason is that sources of light in the office, especially overhead lamps, are so widely spread out that rotating the screen to eliminate reflections from one source may create new reflections from other sources.

Usually, tilting the screen downward can eliminate a lot of reflection from overhead lights. You can also fit a hood over the top of the screen to prevent some sources of light from striking it.

Apart from the wide distribution of overhead and window lights, another reason screen reflections are difficult to eliminate is that the surfaces of the most popular types of screen are slightly convex. A convex surface captures light from a broader area than a flat surface. The convex screen reflects light from a wider area in your work room than a flat surface would. Flat screens are more expensive and, consequently, less popular. The right side-view mirror of your car gives you a wider view of traffic. This is because it is convex like the surface of most computer screens.

Other Surface Reflections

Reflections from desk tops and other furniture, the floor, or walls can be reduced by avoiding the use of highly polished furniture, especially that with polished leather. They can also be reduced by painting walls with low reflective paint and by covering floors with carpets. Light reflected from highly polished furniture is sometimes bright enough to be reflected a second time by the screen. Tiled floors are kept polished for an attractive appearance, but this also makes them highly reflective.

Color and Surface Finish of Keys

The keyboard may provide another surface from which light can be reflected. Inexperienced keyboard users are most susceptible to this type of reflection because they tend to look at the keys frequently. Many experienced users also look at function keys frequently. The keys on older keyboards were smooth and black and reflected a considerable amount of light. Their surfaces were also curved inwards (concave) to conform to the contours of the fingertips. This curvature caused light reflections from many different directions, and some reflections were always in your field of vision when you looked at the keyboard. People tended to tilt their heads to the side to avoid these reflections but were seldom successful. On modern keyboards, this problem has been dealt with by using a surface finish and color that reflects little light on the keys. The surface curvature on the keys, however, has been retained because it conforms to the contour of the fingertip. The reflections have not been eliminated, but they have been reduced to tolerable levels.

Antiglare Filters

An antiglare filter is a flat glass screen that can be placed over the computer screen to reduce the amount of light entering your eyes. The light comes from two sources: (1) external light from windows, overhead lamps, and reflections from certain surfaces hitting the screen and then your eyes, and (2) light generated by the screen characters themselves.

The filter reduces screen reflections from external light by partially blocking out light twice—when the light is about to reach the screen and when it leaves the screen after being reflected off of it. The light from the characters is reduced when it comes off the screen. In other words, reflected light is reduced more than the characters' light. This is good because it enhances contrast and easy viewing. If the light from the characters is reduced too much, you can brighten them by simply adjusting the brightness knob on the screen.

Antiglare filters have been found to be useful, but they are far from perfect. Their effectiveness is sometimes exaggerated by vendors and manufacturers, who often claim that their filter can eliminate reflections. But filters do not eliminate reflections from bright sources. They merely reduce their brightness. If you cover your screen with one of these filters, you can still see reflections from overhead bulbs and windows, though with reduced brightness. And the reduction in brightness may not always be great enough for you to read with ease.

There are three main types of filters—mesh screens, diffusing (light scattering) filters, and optically coated glass filters. They are not all equally effective. Mesh screens have a fine mesh in them, and they are probably the least effective kind. They also restrict your viewing angle. You don't see well from the side of the screen. They also decrease the sharpness of the image that you see. Diffusing filters made of a chemically etched glass or plastic surface that scatters reflected light work better. Optically coated glass filters use a special coating on a glass. They are the best of the three.

Difficulties for Older People

Supervisors should be aware of the limitations and special lighting needs of older workers. Your vision system does not work as well in middle age and beyond as it did in youth. Your visual acuity decreases about 25% from twenty-five to sixty years of age. As you age, your eye lenses gradually become less translucent because of the death of cells at their center. These are the oldest cells in the lens, and they become more separated from the blood system as more and more cells are added to the lens during growth. Lacking oxygen and food substances that are delivered by blood, these central cells gradually die and harden, making the lens less translucent and less flexible. Since they block some light from passing through, you will need brighter light to see as well as you did when you were in your twenties. Scientific studies have established that, on the average, a fifty-year-old person requires a little more than 1½ times the brightness of light as a twenty to twenty-five year old, and a sixty-five year old almost 2¾ times as much to work in an office!

In the computer workplace you can use a task lamp to supplement overhead or ambient lighting. Increasing only the overhead lighting can create problems where younger and older people work in proximity. What is just bright enough for an older worker may be too bright for a younger one.

Since the eye lens becomes more inelastic with age, older people change focus more slowly. This causes their work performance to deteriorate if they are required to look from one visual display to the other rapidly, for example, from the screen to a document and vice versa. The situation is worse if the displays are at different distances from the eye, and they usually are. This will force older people to work more slowly. Otherwise, they will make more errors. However, smart supervisors can avoid this problem if they assign such workers to jobs that require less frequent change of vision.

Eyeball Rotation and Muscle Fatigue

The eyeball is able to rotate within its socket in all directions by the six extrinsic eye muscles (see Figure 19). Each muscle moves the eye in one specific direction, but they all work together during movement to provide smooth, coordinated rotation. Two of these muscles, the **superior rectus** and the **inferior oblique**, can rotate the eyes upward and back to the relaxed midposition. Two more, the **inferior rectus** and the **superior oblique**, can rotate it downward and back. The **lateral rectus** can rotate it outward, toward your side, and back. And the sixth one, the **medial rectus**, can rotate it inward, toward the center line of your body, and back to midposition.

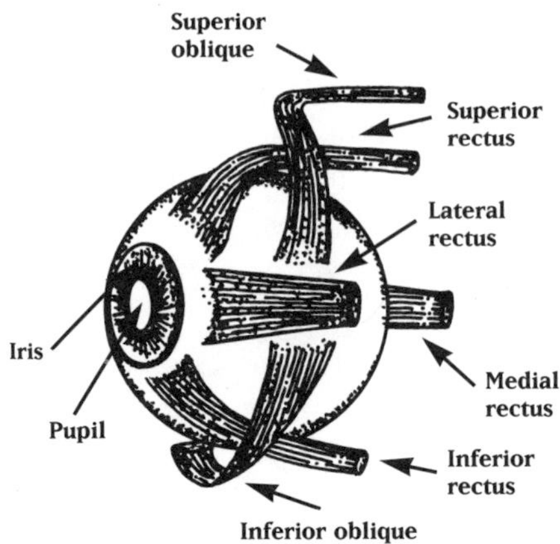

Figure 19. The oculomotor muscles, which rotate the eyeballs.

When the eye is looking straight ahead at a distance, the tension in these six muscles is balanced, and their overall tension is at a minimum compared to any other position. The muscles are then said to be in a relaxed state. But when the eyes must be kept in any rotated position, the muscles that are responsible for the rotation in that direction are in a state of heightened tension. After prolonged periods, these muscles may easily become fatigued, and you may eventually suffer from aches in your eyes or head. For example, the closer an object is to your eye, the greater the inward rotation of the eyeballs must be and the greater the tension in the medial rectus muscle, which controls inward rotation, will be. In computer work, your eyes perform close work all day because they are always fixed on objects that are close to them—the computer screen, the keyboard, or documents. To relieve this tension, during your break periods you should take the opportunity to look at objects at a distance so that the muscles get exercise.

Certain conditions in computer work may also cause you to rotate the eyes away from the midline, increasing the tension in some of the extrinsic muscles. These conditions include the following:

- Documents not being located in a central position in front of you.
- Reflection glare from the screen.
- Reading documents through the bottom lens of your bifocal glasses.

Maintaining your eyes in extreme rotation will involve great tension in the muscles concerned, and you will feel the effects of the tension as a tightness near the eyeballs. Few people will attempt to maintain a sharp rotation for more than a few minutes, and some people can't tolerate it for more than a few seconds. But a lot of aches from eye-muscle work come from the accumulation of mild strains from moderate rotation over a long period of time. These strains may be so mild that you never really feel the muscles tightening. To prevent these strains, you should apply the advice given in this chapter and eliminate conditions that make your reading difficult.

POINTS TO REMEMBER

- ❖ Visual strain is caused by defective vision or intensive reading under poor lighting conditions—excessively bright light (direct glare), dim light, or reflections from smooth or shiny surfaces (reflection glare).
- ❖ Visual strain is manifested as inability to focus, aches in the eyes or head, irritation and watering of the eyes, or burning or soreness in the eyes, etc.
- ❖ Musculoskeletal aches and pains in the back of the neck occur when displays (such as the screen or documents) are located too far sideways and you are forced to twist your neck to read from them, when you try to avoid glare by twisting your neck, or when you tilt your head backward to read the screen through the reading lenses of your bifocals.
- ❖ Sit with windows at your side to minimize glare.
- ❖ Try orienting your screen sideways or vertically to get screen reflections out of your eyes.
- ❖ Use a hood over the screen to help block out some unwanted light.
- ❖ Use a filter screen over the computer screen to reduce screen reflections. But be careful because some loss of sharpness of the characters on the screen may occur, especially with poorer quality filters, and this can contribute to eyestrain.
- ❖ Avoid highly polished or light-colored surfaces in your workplace (desk tops, leather chairs, the floor, etc.). They reflect too much light.

CHAPTER 15

VISION DEFECTS AND EYESTRAIN

A vision defect represents a condition in your eyes, optic nerve, or the visual center of your brain that prevents you from seeing normally. There are many kinds of vision defects. Many, such as elongated or shortened eyeballs, can easily be corrected with glasses or contact lenses. Defects in vision may be due mainly to natural growth from birth, poor nutrition, or intensive and prolonged visual work in a poorly illuminated environment. In the computer workplace, eyestrain is usually the result of intensive visual work, often under lighting that may be too dim, excessive, or uneven. But some cases are partly caused by or exacerbated by defective vision. The defect need not be great for eyestrain to occur—the long duration and intensity of computer work may be enough.

Many vision defects involve blurred images, the inability to focus objects sharply on the retina. Glasses are one of the most popular corrective devices. However, wearing glasses does not necessarily guarantee that you will be free of visual problems. Glasses themselves can pose problems in the computer workplace that are uncommon at most other workplaces. They restrict your depth and range of vision, and they can force your neck and head into constrained and stressful postures. Since the screen, documents, and keyboard are usually positioned at different distances and cover a wide area, you are forced to refocus or move your head or eyeballs constantly during work. Your head and neck posture may also be affected, leading to musculoskeletal strain, especially in the neck-shoulder region of the body.

Common Types of Vision Defects

The difficulties you may experience at the computer workstation depend on the particular type of defect you have, especially if you suffer from more than one type of defect. The most common defects are near-sighted vision (myopia), far-sighted vision (hyperopia), presbyopia, astigmatism, and phoria.

Because of the tremendous strain that computer work imposes on your eyes, you should check them regularly at your optometrist's office. Many eye specialists recommend testing twice a year.

NEAR-SIGHTED VISION (MYOPIA)

Near-sighted people see close objects well but not distant objects. How well you can see is measured in relation to a person with excellent vision. If you see at twenty feet as well as a person with normal vision sees at the same twenty feet, then your vision is said to be normal or 20/20.

It seems as if constant short-distance viewing over a period of time may be responsible for myopia. The populations in the Western and industrialized Far Eastern countries have a greater prevalence of myopia than the populations in nonindustrialized countries. Some experts believe that this is because people in industrialized countries perform more close-up work throughout their lives. This means the eye muscles that are responsible for long-distance viewing do not get enough exercise. These muscles may lose some ability to contract and relax properly. Computer workers should occasionally exercise these long-distance muscles by frequently looking at far-away objects for a few seconds. But this can only be done easily if there is distance within your view, as in working near to a window or in an open-plan office. In isolated cubicles or small offices, this is impossible.

Near-sighted vision is corrected by wearing glasses or contact lenses. Approximately 36% of Americans require prescription lenses to correct myopia and achieve 20/20 vision.

If you are near-sighted and do not wear corrective lenses, the characters on one or more of the displays (screen, keyboard, and documents) may be blurred at their typical distance. They would be too far away from your eyes. You would have to reduce the distance between your eyes and the display. The three main methods people use are thrusting the head and neck forward, with little movement of the trunk, keeping the head and neck fairly stable but thrusting the whole upper body forward, or a combination of these two.

The exact strategy used depends on the distance of the display from your seat and on your personal habits. In all three cases, a part of the body is thrust forward, increasing stress and strain in the neck or lower back. To maintain your bent posture, the muscles, tendons, and ligaments in the neck and lower back must maintain a high state of tension, as described in chapter 8. The spinal discs are also constantly compressed from the increased muscular load. If this posture is maintained for long periods, fatigue and pain can develop in the back of the neck and the lower back.

These sources of musculoskeletal strains can be avoided if you have your eyes checked regularly and wear corrective lenses when you need them. This will prevent a lot of unnecessary bending and leaning. If, after wearing corrective lenses, the screen, keyboard, or document is still out of your range of focus, then the display should be moved closer to you or you should move your chair closer to it. Avoid bending your neck and trunk forward.

Sometimes it is impractical to move the computer screen closer to you. Move your chair forward instead, but not too close, or your elbows will be forced backward, creating new problems. Also, if your desk is not high enough, sitting close to it may cause your thighs to be squeezed by the underside of the desk.

Some people try to compensate for uncorrected eye defects and glare by squinting their eyes to focus properly. However, squinting requires tension in the muscles in the face around the eyes, and this tension can lead to headaches.

FAR-SIGHTED VISION (HYPEROPIA)

Far-sighted vision (also called **hypermetropia** or **hyperopia**) is the condition in which you see distant objects well but not close ones. Reading lenses are generally used to correct hyperopia, but they also restrict your depth of focus. With reading lenses, you must bring the display closer to your eyes than normal, and this can cause other problems.

If you work with uncorrected far-sighted vision, you will be forced to move your eyes away from the screen, documents, and keyboard for focusing. You may either move your chair backward or tilt your head backward. In either of these two cases, your body will be subjected to increased mechanical stress. With your chair moved backward, you may be forced to extend your arms to reach the keyboard, and this requires extra work from the shoulder muscles. With your head tilted backward, excessive stress may be placed on the neck muscles, ligaments, and joints. With your eyes at a distance from the screen, you may be able to bring the material on the screen to a sharp focus, but it may look too small for easy reading.

PRESBYOPIA

As people get older, their ability to see up close declines, and they tend to position displays further away from their eyes. This condition, which becomes noticeable at about 40 to 45 years of age, is called **presbyopia**. It is due to the loss of elasticity of the eye lens and its ability to change shape to focus on close objects. The closest distance from the eyes at which you can focus an object is called your **near point**. With presbyopia, your near point gradually moves further away from your eyes. To appreciate the drastic effect of age on the near point, examine the figures below.

Age	**Near point**
16 years	3 inches
32 years	4¾ inches
44 years	8 inches
50 years	20 inches
60 years	40 inches

If you fail to wear corrective lenses, the ciliary muscles will try to compensate by contracting more strongly, and given the long period of time during which this would occur in computer work, even slight presbyopia may lead to eyestrain.

ASTIGMATISM

If the surface of the cornea of your eye is uneven, light falling on the retina will be scattered, and the image in your eyes will be blurred. This condition is called **astigmatism**. Computer users with astigmatism tend to tilt their heads to one side to minimize the distortion. The neck muscles that control side-to-side movements of the head are then under greater tension than normal.

PHORIAS

To see properly, the image in each eye must be in exactly the same relative position on the retina. This is called **convergence** of the eyes. If the strengths of the muscles that move the eyeballs are not balanced, then the corresponding images from the two eyes will not converge perfectly and you will see double. This is called **phorias**. Without corrective lenses, the muscles can easily be strained in their efforts to compensate and force convergence.

Range of Vision

Ideally the screen, keyboard, and documents should be within your range of vision to avoid any unnecessary bending. With 20/20 eyesight, your range of vision extends from a few inches from the eyes to miles in front of you. For reading, this range extends from a few inches to about three feet. However, when you wear strong corrective lenses for hyperopia, this range becomes shorter. Strong reading glasses cover a range of about 11 to 16 inches from the eyes, and weak ones about 26 to 79 inches. Typical reading glasses for a fifty-year-old person cover a range from 16 to 26 inches.

The keyboard sits in front of the screen, and documents usually lie or stand at the side and a little forward. Research has shown that computer users tend to sit with their eyes at about 24–36 inches from the screen and at about 18–20 inches from the keyboard. This arrangement could easily put one of these visual displays out of focus when the other is within focus if you try to read only with your reading lenses. Some people do this. If you are a typical fifty-year-old using a typical computer arrangement, all three of these visual displays may fall beyond your range of focus. You may then be tempted to bend forward to get your eyes closer to the desk. Your neck and back will then be forced to do what healthy eyes or corrective lenses should be doing.

Area of Vision

Many people over forty years of age suffer from both myopia and hyperopia so they wear bifocal lenses. These lenses compound postural problems associated with computer work.

A bifocal lens is simply two different types of lenses in one frame, one above the other. The lower lens is used for close viewing while the upper lens is used for distance viewing. Some bifocal wearers find it easier to read their screen through their reading lenses. Unless the screen is as much as 45 degrees below the horizontal (which is rarely the case), this requires bending the neck backward. This position tires the neck muscles and joints fairly quickly. It is one of the main causes of musculoskeletal fatigue at the base of the neck among wearers of bifocal and trifocal glasses.

Reading the screen through the top (distance) lens can prevent neck extension and put the neck and head in a more comfortable position. The screen must be positioned at some distance for focusing of the eyes to occur, and this distance may be too great for the characters to appear large enough to be read easily. This problem was worse many years ago when the smaller twelve-inch screen with smaller characters was popular. Today the fourteen-inch screen with larger characters is more common. Even larger screens are available from computer vendors, but they are more expensive than the fourteen-inch ones.

Line of Sight

The distance between your eyes and the visual display is not the only factor that influences the mechanical stress in your spine. The height at which the screen is positioned is also important.

With a high screen you are forced to bend your neck backward, and with a low screen you are forced to bend your neck downward. The consequence in either case is neck strain—tensed muscles and compressed spinal discs. There is a certain position of the neck that imparts minimal stress on its associated muscles and joints. Scientific experiments have shown that this position varies widely from person to person but that, for most people, it is with the neck bent slightly downward. In this position you look along a path below the horizontal at eye level. Ergonomists call this path your normal line of sight.

Scientific studies have shown that when you sit your normal line of sight is 10–15 degrees below the horizontal. Considering that the eyes can rotate 15 degrees upward or downward comfortably, this gives you a field of vision from 5 degrees above the horizontal to 45 degrees below it. Position your computer screen within this range, with its center preferably 10–15 degrees below the horizontal. The height adjustability (or lack of it) of your desk and chair will determine the limits of this range. Height-adjustable furniture gives you more scope for setting your screen height. Experiment with your equipment. Change the screen height on different days. After a time, you should be able to determine the height that is most comfortable for your neck, back, and eyes.

Many years ago, it was widely believed that the screen center should be horizontal with your eye. This imparts an upright posture to your head and neck. Today we know that this is uncomfortable for the majority of people.

If you place your screen flat on the desk top instead of on the computer unit, it will be several inches lower, and this will help get its center below the horizontal at the eye. Unfortunately many people place their screens on the computer unit. In many cases, this brings the center of the screen at or above eye level. The effect is extra work on the neck muscles and joints to maintain the head upright or tilted backward. Some desk designs also make it difficult to position the screen at the correct height. Pull-out keyboards are usually several inches below the desk top and limit how high you can sit. This may cause the screen to be a little too high. Using a desk with a cut-out section on its top will get the screen below the horizontal. The screen sits on a stand below the desk top and inside the cut-out section. The stand may also be adjustable in height.

Frequent Change of Focus

If you wear bifocal lenses, you are likely to experience some peculiar difficulties when working at the computer. Every time you change your view from one of the lenses to the other, your eye lens must also change focus. Inexperienced keyboard users change focus frequently from one type of display to another, for example, from the screen to the keyboard and vice versa. Some types of jobs also require workers to change focus frequently. Ideally, the eyes should change focus instantaneously, but people who wear bifocals or trifocals tend to change focus more slowly. They have two types of problems. One has to do with aging and the other with the design of bi- and trifocal glasses. If accommodation takes half a second, it may tire the eyes when done frequently and rapidly. By age fifty years, accommodation takes about a full second. This is too slow for fast-paced computer jobs. The eyes become easily fatigued as they struggle to refocus quickly to keep pace with work demands.

If you wear bifocals and your eyes accommodate slowly, you may be tempted to read from the screen, keyboard, and documents through only one of the two corrective lenses to avoid the refocusing problem. If you use only the lower lens of your glasses, you will be forced to tilt your head backward considerably to read from the screen or documents and impose excessive load on the muscles and joints of the neck. The higher your screen or documents are, the farther you must tilt your head. The keyboard is usually low enough to prevent tilting of the head when reading through the reading lens. If you use only your top lens for viewing the displays, then you will be forced to move backward to bring the displays within your range of focus. This will make it necessary for you to extend your arms to reach the keyboard, with consequent strain in the shoulders. Trifocal wearers experience similar problems.

Computer Glasses

The problems of changing focus or bending your neck associated with bifocal or trifocal glasses can be overcome by the use of special eyeglasses popularly known as computer eyeglasses. These are glasses with a single lens that has a wide depth of focus, covering the distance between the screen, keyboard, and document. They are like reading glasses with a deep range of focus. Unfortunately, these glasses create another problem—your range of vision is severely limited. Objects beyond your computer screen are not sharply focused. Just looking at something a few feet away may become bothersome. This restricted range of vision may make you feel as if you are in a small world. Some employers may be happy about this since they may expect you to be visually undistracted by things beyond your immediate workstation. But minor interruptions are not always harmful. They may be therapeutic for boredom or for speed and load stress.

Another possible problem with computer glasses is that your eyes are forced to maintain a narrow range of focus throughout the workday. The focus-regulating ciliary muscles may then be deprived of the exercise that keeps them healthy.

POINTS TO REMEMBER

- Defects in your vision can cause visual strain from computer work.
- All computer users should have their eyes tested regularly by an optometrist and should wear corrective lenses at all times if they are prescribed. Working at the computer without your corrective lenses can lead to severe eyestrain.
- The ability to change focus quickly is called accommodation. At about forty to forty-five years of age, your eye lenses become less elastic and you begin to lose the power of accommodation, making jobs that require frequent changes in focus from the screen to documents more difficult.
- Myopia (near-sighted vision) is the difficulty in focusing on distant objects. Uncorrected myopia can cause you to bend forward at the neck or trunk to bring your eyes closer to the screen or documents.
- Hyperopia (far-sighted vision) is the difficulty in focusing on near objects. Uncorrected hyperopia can cause you to sit back to get the screen sufficiently far from your eyes.
- Astigmatism is the inability to focus because of unevenness of the front surface of the eyeball (the cornea). It causes you to bend your head sideways.
- Phorias is the inability to focus the image of an object at the same relative spot on the retinas of both eyes. Uncorrected phorias can cause fatigue in the muscles that rotate the eyeballs.
- If you try to read the computer screen from the lower lens of your bifocals, you must raise your head, leading to neck strain.
- Special computer glasses can eliminate neck-bending problems for bifocal or trifocal wearers, but they also limit the range of vision you need when looking away from your workstation.
- Most people can minimize neck-muscular problems by placing the center of the screen 10–15 degrees below the horizontal.
- Exercise your eye muscles by alternately looking at near and far objects during brief breaks from work. It will help maintain your eyes' ability to change focus quickly and prevent early myopia problems.

CHAPTER 16

REST BREAKS AND EXERCISES

A rest break at work is a period of time during which you cease working so that your body and mind can recuperate from fatigue and so that you can attend to other physiological needs. The official rest breaks in most U.S. workplaces are two ten- to fifteen-minute coffee breaks and a thirty-minute meal break, interspersed at equal intervals of time. These breaks were established well before the advent of computer jobs. Although they are generally accepted without question, they do not necessarily reflect the amount of time needed by workers in certain jobs to recuperate from fatigue. They are based on convenience rather than objective scientific knowledge of how our bodies work.

In precomputer age office work, the traditional 15-, 30-, and 15-minute breaks worked because of the varied nature of tasks. Traditional office work provided typists, for example, with many short periods during which they stopped keying on the keyboard and performed other tasks instead, tasks that used other muscles in the body. They changed paper on the typewriter, used correction fluid, retrieved documents from files, and so on. This type of work provided adequate minibreaks, which gave typists the opportunity to rest the hardest working muscles while others took over. The muscles that controlled the hands over the keyboard and those in the back, neck, and shoulders received much needed rest when work was varied in this way. Even though some of these unscheduled minibreaks (micropauses) lasted less than a minute, their accumulated effects over the workday definitely helped delay the onset of muscular or mental fatigue. (A micropause is considered a pause from work lasting from about 2 to 60 seconds.) But the strict organization and control of many types of jobs in the computerized office have eliminated these rejuvenating micropauses.

The typical computer user is deprived of this variation of tasks because corrections can be made on the computer, files can be retrieved on the computer, the printer can be set on the computer, and so on.

Employers treat computer jobs as if they were traditional secretarial jobs, especially with respect to rest breaks. This is wrong and reflects a lack of understanding of the human body at work. The intensity of some types of computer work and their physical and psychological effects on the body clearly indicate that people who work at computers need more rest breaks than office workers of twenty years ago. Perhaps the term rest break affects our perception of things. It sounds like a gift from employers, and it may create an image of a person in a perfectly relaxed state. But this is not necessarily so for computer workers. Perhaps the term "recuperation break" or "recovery break" should be used more often. That sounds more like reality. It reflects the needs of the worker—the needs of the physical body and the mind.

There is a limit to the time you can perform work at a certain level of intensity without rest breaks. Ergonomists call it the **endurance time** for that job. For a typical data processing job at the computer, this time is likely to be less than an hour. If you work without rest beyond endurance time, your body fatigues and you make more errors than normal. You can maintain accuracy if you reduce your work rate. If you don't, your body will rebel. Fatigue turns into aches and pains, and you develop anxiety, irritability, and a host of other psychosomatic reactions. However, short, frequent pauses can reverse many of these work-induced health disorders.

Office computer work is relatively new, and few people know exactly how much rest is needed by people who work at stressful computer-oriented jobs. In this competitive age, management's strategy has been to allow no more break time than is necessary for maintaining a desired level of productivity. Usually, these break times are the same as those common in other workplaces or the maximum amount agreed upon by management and unions. They are the typical 15-, 30-, and 15-minute breaks. The fact that jobs vary widely in their degree of stress is rarely considered. The frequency and severity of health complaints from computer workers indicate clearly that the normal office break times are inadequate for many computer jobs.

Even concerned employers who are willing to allow more break times have no reliable body of scientific data to guide them. Scientific studies for establishing work breaks for any type of work are few and inconclusive. The few studies that have been done are mostly for jobs in which some observable change in the human body can be accurately measured. Energy exchange in the body is one such change. For example, ergonomists can evaluate the strain for shoveling jobs in coal mines by calculating the amount of kilocalories of human energy required to perform the job. We measure the amount of oxygen consumed by the person at work and convert it to energy produced by the body. (One liter of oxygen taken into the body produces approximately 5 kilocalories of energy.) We also measure the person's maximum capacity for producing energy. By comparing job energy requirements and a person's energy capacity we can then evaluate the severity of the job—more severe jobs require a greater percentage of a person's capacity and less severe jobs a smaller percentage. This data can then be used to determine work-break schedules.

Unfortunately, for computer jobs there is no such measurable physiological or mental change that accurately reflects your body's strain. This seems to have influenced the retention of the traditional, but inappropriate, office work-break schedule. However, the level of complaints and evidence of musculoskeletal problems among office computer workers suggest that computer work is much more stressful than traditional office work and requires a different work-break schedule.

Overtime and Extra Hands

Employers are often worried that rest-break times above the normal amount can be expensive for them. They tend to think that they would have to compensate for this lost time by hiring additional employees or allowing overtime, both of which would increase their operating costs. But this is not necessarily true. There are sound reasons to believe that the gains in health and efficiency from increased rest breaks can reduce the need for overtime work. One recent study in San Francisco found that these gains not only offset any perceived loss but also completely eliminated overtime work that was necessary before the extra breaks were instituted.

Overtime work can be disastrous to the health of computer operators. It is not hard to see that computer work, which can hardly be sustained for eight hours comfortably, cannot be maintained for an extra two, four, or six hours without serious repercussions. Remember that musculoskeletal strain is cumulative. This means, among other things, that the amount of strain necessary to harm your body after eight hours of work would be less than the amount required at the beginning of the day. Two hours of extra work at the end of a day can feel the same as four hours of work at the beginning of the day. As the workday progresses, your body progressively becomes more vulnerable to the effects of mechanical and mental stresses. But, even when employees are experiencing the effects of fatigue, many of them may not refuse overtime work because they need the extra earnings. Because of heightened motivation to earn more, you may be able to maintain an acceptable quality of work during overtime. You may also shrug off signs of physiological and mental fatigue for a time. But eventually the accumulated strains will take their toll on your body.

Many employers prefer to hire their regular employees for overtime work instead of hiring extra operators. The employers save themselves some operating costs, and employees increase their earnings. But both seem to be unaware of the harm that overtime work can cause. Using overtime labor eliminates overhead expenses like training or familiarizing a new employee with a particular job. It also eliminates benefits such as health insurance, life insurance, pension funds, and sick leave or regular annual leave that must be granted to extra employees.

How Should Break Times Be Divided?

Some experts have recommended rest-break times of fifteen minutes every two hours, while others have recommended fifteen minutes every hour. And these recommendations are being implemented in some computer workplaces. For example, a recent law in San Francisco, California (later overturned), stipulated that computer operators should be given a fifteen-minute break every hour. Other experts have turned toward an older, more natural method and are claiming greater success in reducing cumulative strains. By this method, employees are allowed frequent, short microbreaks at their discretion in addition to the traditional office break times. These microbreaks are no different from the impromptu rest breaks that workers in physically strenuous jobs have been allowed in the past, except that they are shorter. Provided that they are not abused, breaks taken naturally are more beneficial than when taken at rigidly scheduled times. Some computer programs can be used to remind you, with distinct beeps, when to take your microbreaks. However, I believe that you know your body better than anyone else and know best when it needs a break.

Scientific studies have also shown that people feel less fatigued and work better when their total break time is spread out over several short breaks rather than over a few long ones. The breaks need not be evenly spread out over the day, provided that they are not bunched up too closely.

The results of one study should serve as an indication of the effectiveness of short, interspersed breaks. Eighty-five data-entry operators were allowed to stretch and exercise their arms, necks, backs, and legs during two five-minute periods during the workday for a year. They were also allowed to take twenty- to thirty-second minibreaks to stretch, flex, and shake the affected muscle groups when they felt muscular discomfort during work. The results were astounding. Workers' compensation claims for musculoskeletal problems within the study period dropped from a maximum of twelve to zero, operators' speed and efficiency increased significantly, productivity increased 25%, and overtime work for meeting deadlines was eliminated. In another study in 1981, data-entry operators were advised to take ten to fifteen seconds for micropauses every ten minutes. At the end of the workday, they reported less than half as much muscular fatigue in the neck and shoulder region than before.

Job Enlargement

In physically stressful manufacturing jobs, a technique called job enlargement has been used for combating physical and mental fatigue. By this method, workers are transferred to a less demanding job after a certain time, usually after four hours, and other workers replace them. However, job enlargement requires a versatile workforce and is not easy to administer in the computerized office. It also requires the operators to be proficient at more than one type of job—a requirement that is not common in this age of specialization. You are much less likely to succeed in today's workplace if you are a jack-of-all-trades and master of none than if you are a master of only one and a stranger to others.

Misinterpretation

The use of rest breaks or physical exercises should not be considered a cure-all for aches and pains caused by computer work. Rest breaks can eliminate or reduce physiological imbalance caused by overexertion, and exercises can enhance your resistance to wear and tear, but neither of these two strategies deal with the causes of cumulative trauma illnesses. We must try to deal with the causes first—that is, design work within people's capabilities. Rest breaks should be used as a supplement for ameliorating work conditions. Ergonomic engineering methods should be the first step in dealing with the causes. The design of furniture, ambient lighting, task lighting, workplace arrangement, and work methods should be dealt with before rest breaks are considered. They can reduce musculoskeletal and mental stress significantly and, in turn, reduce the need for extended or frequent rest breaks.

Exercise Regimens

Exercise regimens are increasingly being advertised and written about in the popular news media as a method of maintaining health in workers. Many employers and insurance companies look favorably toward them. But they are not necessarily as useful in computer work as in other types of work. Exercises will enhance your general physiological strength and endurance and your psychological well being, but they will not necessarily prevent the accumulation of strains in muscles and joints from a poorly designed job. Exercises make your body more resistant to physical stress, but they do nothing to remove the sources of the stress. They also inadvertently lead you to believe that the aches and pains you feel from your job are your fault—that your body is weak or unfit. This tends to divert your attention from the real causes of your pains—poorly designed work and work environments.

You should also be careful of the appropriateness of the exercises recommended to you. Exercises that are beneficial for physically fit people may be harmful to those not accustomed to exercising. A recent study funded by NIOSH examined 127 exercises from various workplace books and pamphlets and found many of them to be inappropriate. Half were considered disruptive to work, 40% hazardous to health, and one-third conspicuous and embarrassing. Some of the hazards included extreme bending of the trunk and wrists, which stretches muscles and tendons that are already overworked and fatigued.

Most exercises fall into one of two categories—muscular and cardiovascular. Muscular fitness exercises involve contractions of your skeletal muscles (those that are responsible for movement and posture) while trying to resist forces, as in weight lifting.
Cardiovascular-fitness exercises involve strengthening the muscles of the heart so that blood can be pumped around the body effectively. In cardiovascular exercises, you move about vigorously enough (as when jogging, doing aerobics, and playing basketball)

to stimulate the muscles to produce extra energy. Blood flow must increase to transport extra food substances and oxygen to the muscles. The heart pumps harder and faster and is gradually strengthened. You can get a lot of cardiovascular exercise by simply walking. Walking up and down stairs is excellent exercise for the heart, legs, and back, provided you are in good health.

Exercise that can best help you combat mechanical strain from computer work is not necessarily of the above types. Helpful exercises include stretching of muscles that maintain prolonged static contractions and bending of joints that are kept still during work. The objective is to stretch and bend as far as you can then return to the original position in a slow rhythmic motion without feeling discomfort. Repeating each exercise a few times strengthens the ligaments, muscles, and tendons around joints and maintains the joints' flexibility. The rhythmic motions also help maintain blood flow in the muscles and around the joints. There are some helpful books in bookstores and libraries with descriptions and diagrams illustrating these bending and stretching exercises. If you experience difficulties while performing these exercises, you should consult a physician before continuing. They can be performed at home and at your computer workstation during microbreaks. Those for the arms, shoulders, and neck can be performed without even getting up from your seat. However, they should not be performed while you are actually experiencing pain from overwork.

Vitamins

Many physicians believe that a deficiency of vitamin B_6 (pyridoxine) may be a cause of carpal tunnel syndrome. This is because tests have shown that some CTS sufferers have below-normal levels of pyridoxine in their bodies. This belief has been reinforced by the observation that CTS occurs more frequently in pregnant women than others, and pyridoxine levels in women are lower during pregnancy. So some physicians help treat CTS problems by recommending daily dosages of vitamin B_6. They also recommend the use of vitamin C because of its anti-inflammatory properties and vitamin E to prevent scarring of tendons and to aid circulation. Proponents of the vitamin treatment theory believe that these vitamins eliminate inflammation of the tendon sheath within the carpal tunnel and remove pressure from the median nerve.

Some reports even make the outright claim that vitamin B_6 supplements can prevent CTS, but there is no scientific proof of this. Vitamins may help treat nonwork-related CTS, but it is unlikely that vitamin treatment can prevent CTS that is caused by excessive finger motions in computer work, especially at a poorly designed workstation and with inadequate rest. It is equally unlikely that vitamin treatment can reverse CTS problems if you continue to perform the work that caused it. Vitamin treatments may be another quick-fix solution to a work-related health problem that takes responsibility off the shoulders of employers and puts it on the suffering employee.

A 1993 publication in the *Journal of the American College of Nutrition* confirmed that vitamin B_6 does have a therapeutic effect on CTS patients, especially for pain and loss of sensation, but that a deficiency of this vitamin in the body may not cause CTS. You must also note that high doses of vitamin B_6 may be toxic to your nerves. Consult your physician for advice when considering the intake of large amounts of this or any vitamin.

POINTS TO REMEMBER

- ❖ Rest breaks help your body to recuperate from muscular and mental fatigue.
- ❖ The traditional 15-, 30-, and 15-minute rest-break schedule is not sufficient for the computer workplace.
- ❖ Rest breaks should be taken whenever your body calls for it or your health may suffer.
- ❖ Several short breaks are better than one that amounts to the same total time.
- ❖ Perform stretching exercises during microbreaks and walk around your office. These exercises enhance your blood flow and delay fatigue.
- ❖ Vitamin treatment is not a cure for work-related CTS.
- ❖ Extra breaks, exercises, vitamins, and the use of a wrist restraining device should not take the place of redesigning your workplace.

AFTERWORD

As easy as it may seem to casual observers, computer work is stressful. The stresses affect you physically and psychologically. They arise from the nature of the work and the physical attributes of the computer workplace—high-speed performance; repetitiveness; poorly designed keyboards, chairs, and desks; poor workplace design; and poor lighting. Stressful body postures and visual strain are common. The results, for many people, are chronic aches and pains that may take weeks, months, or even years to manifest themselves.

People who deny that computer work can produce aches and pains are not being realistic. Many employers who were skeptical in the past have now changed their minds. Knowledge about aches and pains in the computer workplace is spreading rapidly thanks to increased research into the problems, more education and training programs, recognition by workers' compensation laws, involvement by employers and insurance companies in controlling the problems, and leadership by OSHA in monitoring and controlling them. Ironically, one of the two greatest driving forces for controlling CTDs in the workplace is the rising cost to employers for medical bills and loss in productivity. The other is the threat of heavy fines by OSHA for violating safety laws. Many employers are now trying to minimize these costs by creating ergonomic programs for workplace safety. Those in the forefront of these efforts are beginning to hire qualified ergonomists to work full time or as consultants on special projects. They are offering education and training to all affected employees, and they are performing timely ergonomics investigations in the workplace.

The recent release of a draft for an ergonomics protection standard by OSHA to control CTDs in the workplace emphasizes the gravity of these problems. OSHA has proposed strict guidelines on how to manage safety from an ergonomics approach. It states that employers must provide ergonomic awareness and job-specific training to all employees and supervisors for jobs that pose a risk of injuries and illnesses. Employers must also have a medical management program, which includes diagnoses, treatments, descriptions of restricted activities, and follow-up examinations for affected workers. To help in these efforts, OSHA has included in its draft detailed information on conducting an ergonomics program, including checklists for workplace investigations. At the completion of this book, the OSHA draft was under review by experts and employers. Its final form is expected to control and reduce CTDs in the workplace to manageable levels.

Although there are many criticisms of the methods OSHA proposed for dealing with workplace CTDs, OSHA has gone a long way in trying to ease suffering among workers. But the real success of these efforts depends on the establishment and execution of effective ergonomics programs by employers. Employees must also cooperate with their employers and become involved in the efforts to control the problems. They must not only make reports and seek treatment and compensation for their illnesses and injuries but also help themselves with the education provided by their employers and demand that education if it has not been provided. You need to know not only what the problems and proposed solutions are but also why the proposed solutions will work. The "why" is as important as the "what" and the "how." As this book has shown, computer users and other workers who sit all day can do a lot for themselves instead of depending on experts to solve work-related problems and get rid of their aches and pains.

GLOSSARY OF TERMS

Accommodation: The change of focus of the eye lens to see an object in proper focus.

Ambient lighting: Lighting in the surrounding environment, usually from overhead lamps.

Anthropometry: The study of dimensions of the body and ranges of motions of body segments.

Antiglare filter: A flat glass plate that can be placed over the desktop-computer screen to reduce glare from light reflected off the screen.

Astigmatism: The eye defect in which the surface of the cornea is uneven and light is scattered on the retina, causing you to see blurred images.

Backrest angle: The angle between the backrest and seat pan of a chair. It controls your hip angle and affects the pressure in your lumbar spine while sitting.

Balans chair: It has no backrest but has a knee rest that allows you to sit upright in a posture about midway between sitting and kneeling. This posture opens the hip angle and helps prevent flattening of the lumbar curve.

Bursa: A flattish filled sac lying next to a joint. It acts as a cushion over which tendons slide and on which they press during muscular contraction. The cushioning effect protects the tendons from mechanical strain.

Bursitis: Inflammation of a bursa. Shoulder bursitis is common. Work-related causes include prolonged or repeated elevation of the arm, which causes strong tension in the tendons of the shoulder muscles.

Carpal bones: The eight small, rounded bones at the back of your wrist. They form the rear wall of the carpal tunnel.

Carpal ligament: The tough band of tissue around the palmar side of the wrist forming part of the carpal tunnel wall. It is also called the transverse carpal ligament because it runs across the wrist.

Carpal tunnel: The channel inside your wrist through which blood vessels, tendons, and the median nerve of the arm pass as they run from the forearm to the hand.

Carpal tunnel syndrome (CTS): A type of cumulative trauma disorder in which swollen tendon sheaths inside the wrist press the nearby median nerve onto the wall of the carpal tunnel, causing extreme pain. Other symptoms of CTS include partial loss of hand strength and restricted hand and wrist motions. Common work-related causes include wrist bending, repetitive motions, muscular contractions, vibrations, and inadequate rest. CTS can be caused by excessive keying in some types of computer jobs.

Cartilage: Elastic and compact connective tissue found at the ends of some bones. It prevents the ends of the bones from excessive wear and tear.

Casters: Wheels on the legs of a chair.

Center of gravity: The position in an object around which the mass seems to be evenly distributed. Theoretically, an object can be balanced on a pivot directed at its center of gravity.

Cervical spine: The part of the spine in the neck region of the body that curves forward. Your neck bends and rotates in this area. Neck pain may be due partly to excessive pressures in the discs in the cervical spine.

Ciliary muscles: The muscles that control the thickness of the eye lens and the focusing of the eyes.

Coccyx: The bottom bone in the spine that curves like a bird's beak, also called the tail of the spine.

Compressive force: A force that squeezes an object in contact with it. An example inside the body is the weight of the upper body squeezing the discs in the spine downward.

Computer glasses: Special single-lens glasses prescribed by an optometrist that extend your depth of vision for reading. Bifocal and trifocal wearers may find such glasses useful for computer work; they will be able to read the keyboard, documents, and screen through this single lens.

Cumulative trauma disorder (CTD): The illness or disorder caused by the gradual accumulation of strains (even weak ones) caused by mechanical stresses over a period of time. The term refers to strains in any part of the body, but it is the upper body, especially the arms and shoulders, that suffers most from CTDs.

Depth of vision: The distance in front of your eyes over which you can focus and see properly. Reading glasses limit your depth of vision significantly. Computer glasses extend the depth so that you can see and read from the keyboard, documents, and screen easily. Depth of vision is also called range of vision.

Desktop computer: The computer that sits on desks in offices and other workplaces. It is the most common type of computer in use today. It is also referred to as a PC (personal computer) or desk computer.

Direct glare: Excessive light entering the eyes directly from a source; for example, when you are facing an open window on a bright day.

Disc: The tough doughnut-shaped compressible structure that separates vertebrae in the spine from one another. They protect the vertebrae from wear and tear.

Document holder: A small stand on which you can place loose documents upright.

Dynamic muscular contraction: Contraction of a muscle so that its length changes and the body part containing it moves, for example, the contraction of the finger muscles when you use a keyboard.

Endurance time: The maximum amount of time a person can work without fatigue.

Epicondylitis (tennis elbow): Inflammation of the tendons at the elbow from excessive work by muscles in the arm. Using the keyboard for prolonged periods can cause tennis elbow in the computer workplace.

Ergonomics: The study of the interaction between people and work and the design of work systems with the goal that people can perform their tasks properly. This is achieved by making the demands of the task within the capacities of people's functions.

Eyestrain: The condition having any of a number of the following symptoms resulting from overwork of the eyes—irritation and watering of the eyes, burning, soreness, inability to focus properly (blurred and double images), seeing colored fringes on objects, headaches, aches in the eyes, etc. In computer work, vision strain can occur from prolonged focusing of the eyes on the screen. The symptoms are temporary and disappear after adequate rest.

Far-sighted vision: The eye defect in which you cannot see near objects clearly, also called hypermetropia, hyperopia, and long-sightedness. Reading glasses correct this defect.

Forearm rest: A pad on which you can rest your forearms when using a keyboard or mouse. It can be placed on the edge of your desk to prevent pressure on your forearms from the desk.

Glare: Excessive light entering the eyes and preventing you from seeing clearly.

Hamstrings: A group of muscles at the back of your thighs that pulls on and turns your pelvis when you sit. Its strong contractions when you are sitting may affect the pressure in the lumbar spine.

Hip angle: The angle between your trunk and thigh.

Hypermetropia: See far-sighted vision.

Hyperopia: See far-sighted vision.

Ischial tuberosities (sitting bones): The projections at the bottom of each half of the pelvis (hip bones) on which the buttocks rest or pivot when you sit.

Job enlargement: Shifting a worker from one job to another that is less stressful to reduce the overall stress during the workday. Rotation periods usually last two to four hours.

Joint: The structure (ligaments and connective tissue) between the ends of two bones that hold the ends of the bones together.

Kyphosis: The backward curves in the spine in the chest and sacrococcygeal regions (the hip and tail regions).

Lactic acid: A poisonous waste substance produced in your muscles when they are overworked. It produces aches and pains.

Ligament: A cordlike structure that connects the ends of different bones to form a joint. Ligaments hold a joint intact and resist sharp bending to prevent dislocations.

Line of sight: The path along which your eyes look when your neck and eyes are relaxed. It varies widely among people but is usually within 10 to 15 degrees below horizontal.

Lordosis: The forward curves in the neck and lower back regions of the spine.

Lumbar pad: A pad used as a lumbar support.

Lumbar spine: The part of the spine at the lower back region of the body that curves forward. Your upper body bends in this area. Lower back pain is partly due to excessive pressures in the discs in the lumbar spine.

Lumbar support: The forward curved projection in the lower-back region of a chair's backrest. Its purpose is to fit snugly into the forward curve of the lower spine, preventing flattening of and consequent increase of pressure in the spine.

Lux: A unit of measurement of the brightness of light. A typical U.S. office is about 400 to 500 lux bright.

Mechanical strain: Gradual damage to the body from the effects of repeated or sustained forces over a period of time, like slow tearing of the tendons in the wrist from repeated contractions and rubbing on the carpal tunnel wall.

Median nerve: One of the three nerves that run to and control the fingers. They originate from a branch of the spinal cord in the neck area and, unlike the other two (the radial and ulnar nerves), they pass through the carpal tunnel in the wrist. The median nerve is easily pressed against the wall of the carpal tunnel from swollen adjacent tendons, giving rise to the dreaded disorder called carpal tunnel syndrome.

Microbreak: (Also called micropause.) A short pause from work, lasting only a few seconds. Microbreaks may be inherent in the task, as in waiting for a computer to respond to an input, or they may be taken deliberately to help ward off fatigue.

Minibreak: A short break lasting up to a few minutes.

Monitor: The computer system unit that contains the screen.

Musculoskeletal disorder: The effects of cumulative strains on some part of the musculoskeletal system affecting its functions.

Musculoskeletal system: The system comprised of skeletal muscles and their tendons, ligaments, joints, bones, cartilage, and associated nerves.

Myopia: See near-sighted vision.

Near-sighted vision: The eye defect in which you cannot see distant objects clearly. Distance glasses correct this defect. Near-sighted vision is also called myopia and short-sightedness.

Near point: In vision, the closest point in front of your eyes at which you can focus properly. This point gets further away as you get older.

Neutral posture: The posture your body will assume if the effect of gravity can be eliminated, as when floating in space. It resembles a horse back rider's posture.

NIOSH: National Institute of Occupational Safety and Health. The federal agency created in 1970 by the OSHAct for researching hazards in U.S. workplaces, publishing information on hazards, and recommending standards to OSHA.

Occupational disorder: An adverse change in the body that develops over a period of time from stressful work conditions, like lower back pains from prolonged sitting. It is also called occupational illness.

Occupational injury: Instantaneous damage to the body by a sudden force, such as a broken bone from a fall.

Oculomotor muscles: The six muscles that move the eyeball. They are attached to the outer surface of the eyeball and to the bones of the eye socket. The oculomotor muscles are also called external eye muscles.

Open hip angle: An angle of over 90 degrees between the trunk and thigh.

OSHA: Occupational Safety and Health Administration, the federal agency created in 1970 by the Walsh-Healy OSHAct to set and enforce standards for safety in U.S. workplaces.

PC: Personal computer (also called desktop computer).

Phorias: The eye defect in which you see double images of an object because the two images are not positioned at the same relative positions on the retina of each eye. It is caused by an imbalance in the strengths of the external eye muscles.

Postural overload: Excessive work by the musculoskeletal system to maintain a particular posture. Symptoms include discomfort, aches, or pains. Postural overload occurs quickly if the posture is awkward and gradually if the posture is not awkward but is maintained for a prolonged period.

Presbyopia: The eye defect in which your ability to see near objects gets worse with age, usually after about age forty.

QWERTY keyboard: The traditional keyboard with the QWERTY keys at the top row from the left.

Reflected glare: Glare caused by light reflected from any surface, for example, light from overhead bulbs reflected off the computer screen into your eyes.

Repetitive task or motion: A task or a series of motions that is performed over and over with a short time for completion, usually less than thirty seconds.

Sacrum: The part of the spine inside the pelvis (hips) between the lumbar spine and coccyx.

Seat depth: The length from the front edge to the rear edge of a seat pan.

Seat pan: The sitting surface of a chair.

Shear force: A force acting in a direction that can cause sliding of one object over another (frictional force). If the force is great enough, the object will slide. Shear forces act perpendicular to compression forces. In your body, shear forces act at the junction of a vertebra and a disc when the spine is bent in any unnatural position.

Skeletal muscles: Muscles attached to the skeleton, often via tendons. Contraction of these muscles produces body motion and maintains posture.

Split keyboard: A keyboard in which the alphabetic keys are split into two sections that are separated by a specific distance. This and other features of the split keyboard help keep the arms close to a position of minimum musculoskeletal stress.

Static-muscular contraction: Contraction of a muscle such that its length does not change and the body part containing it does not move; for example, the contraction of the back muscles when you sit still.

Strain: The body's reaction to stress.

Stress: An undesirable condition that affects your body adversely. Stress causes strain.

Task lighting: Lighting from a portable lamp that is brought near your work. Some people use task lighting when the overhead lighting is not bright enough.

Taylorism: The method of work developed by Frederick Winslow Taylor and the Gilbreths at the turn of the 20th century in which unnecessary and wasteful motions are eliminated after analyzing the tasks.

Tendinitis: Inflammation of a tendon in the body. Repetitive and prolonged keying on the keyboard is believed to be a work-related cause of tendinitis in the wrist.

Tendon: A cordlike structure at the end of a muscle that connects a muscle to a bone. Tendons become tense when their muscles contract.

Tendon sheath: A tough, fluid-filled covering around part of a tendon where it passes near bony projections such as joints. The tendon sheath protects the tendon from wear and tear caused by constant rubbing and pressure of the tendon on projections.

Tenosynovitis: Inflammation of a tendon sheath or tendon within the sheath. Repetitive and prolonged keying is believed to be a work-related cause of tenosynovitis in the wrist.

Thoracic spine: The part of the spine in the chest region of the body that curves towards the rear.

Torsion force: A force created when one body rotates or tries to rotate about another. When you sit and turn your trunk, torsion forces are created in the spine.

Upper body: The head, neck, trunk, and arms. It accounts for 65% of your body weight.

Vitamin B_6 treatment: The use of high doses of vitamin B_6 in an attempt to prevent carpal tunnel syndrome. There is no evidence that this treatment can prevent work-related carpal tunnel syndrome.

Waterfall seat: A seat whose front edge is curved downward like the top of a waterfall.

Work design: The planning of work to enhance performance.

Workers' compensation benefits: Benefits (money, medical treatment, etc.) stipulated by a state that a worker must be given if she has a job-related injury or illness, regardless of who is at fault.

Work hardening: A form of training for work in which the person performs simulations of the work so as to eliminate or minimize strains and injuries on the job. Work hardening is performed on workers recovering from work-related injuries or illnesses in special work-hardening clinics to get them ready to return to work and prevent recurrence of the injuries. Work hardening is no substitute for prudent ergonomic design of work.

Workplace design: The arrangement of work equipment and furniture in a specific way for facilitating job performance.

Work-rest schedule: The arrangement of work and rest times during a workday. Most experts consider the typical 15-, 30-, and 15- minute rest breaks within the workday inadequate for some types of computer work.

Workstation: The immediate area of your work. The computer workstation consists of a desk, chair, the computer, and other work materials such as a printer.

Wrist rest: A pad placed in front of the keyboard for resting the wrist. Some people use a wrist rest while keying.

Wrist restraining device: A device, such as a splint or strap, used for restricting movement of the wrist. It is useful when recovering from a wrist injury or disorder.

Wrist splint: A rigid device used to immobilize the wrist joint, usually for promoting healing and reducing pain. Contrary to some claims, wrist splints do not prevent cumulative trauma of the hands or any other part of the body.

Wrist strap: A stiff strap worn around the wrist to partially immobilize the wrist joint, usually for promoting healing and reducing pain. Contrary to some claims, wrist straps do not prevent cumulative trauma of the hands or any other part of the body.

BIBLIOGRAPHY

Aghazadeh, F., ed. *Trends in Ergonomics/Human Factors* V, 1988. Amsterdam: Elsevier Science Publishers, 1988.

Andersson, G.B.J., and R. Ortengren. "Lumbar Disc Pressure and Myoelectric Back Muscle Activity During Sitting II: Studies on an Office Chair." *Scandinavian Journal of Rehabilitative Medicine* 3 (1974): 115–21.

Andersson, G.B.J., R.W. Murphy, R. Ortengren, and A. L. Nachemson. "The Influence of Backrest Inclination and Lumbar Support on the Lumbar Lordosis in Sitting." *Spine* 4 (1979): 52–58.

Aoyama, H., H. Ohara, Y. Oze, and T. Itani. "Recent Trends in Research on Occupational Cervicobrachial Disorder." *Journal of Human Ergology* 8 (1979): 39–45.

Armstrong, T.J., and B.B. Chaffin. "Carpal Tunnel Syndrome and Selected Personal Attributes." *Journal of Occupational Medicine* 21 (1979): 481–86.

Armstrong, T.J., L.J. Fine, S.A. Goldstein, Y.R. Lifshitz, and B.A. Silverstein. "Ergonomics Considerations in Hand and Wrist Tendonitis." *Journal of Hand Surgery* 12A (1987): 830–37.

Armstrong, T.J., J.A. Foulke, M.J. Bernard, J. Gerson, and D.M. Rempel. "Investigation of Applied Forces in Alphanumeric Keyboard Work." *American Industrial Hygiene Association Journal* 55 (1994): 30–35.

Arndt, R. "Working Postures and Musculoskeletal Problems of Video Display Terminal Operators—A Review and Reappraisal." *American Industrial Hygiene Association Journal* 44 (1983): 437–46.

Awerbuch, M. "RSI, or 'Kangaroo paw.'" *Medical Journal of Australia* 142 (1985): 237–38.

Ayoub, M.M. "Workplace Design and Posture." Human Factors 15, no. 3 (1973): 265–68.

Ayoub, M.M., H.J. MacKenzie, and S. Deivanayagam. *Training Manual in Occupational Ergonomics*. Lubbock, Texas: Institute of Biotechnology, Texas Tech Univ., 1976.

Bammer, G., and I. Blignault. "A Review of Research on Repetitive Strain Injuries (RSI)." *Proceedings of the Australian National University Research on RSI*. (1986) 5–9.

Bammer, G. "Musculoskeletal Problems Associated with VDU Use at the Australian National University—A Case Study of Changes in Work Practices." In *Trends in Ergonomics/Human Factors III*, edited by W. Karwowski, 285–93. Amsterdam: Elsevier Science Publishers, 1986.

Barkla, D.M. "Chair Angles, Duration of Sitting, and Comfort Ratings." *Ergonomics* 7 (1964): 297.

Bendix, T., J. Winkel, and F. Jenssen. "Comparison of Office Chairs with Fixed Forward or Backward Inclining, or Tiltable Seats." *European Journal of Applied Physiology* 54 (1985): 378–85.

Bendix, T., and F. Jenssen. "Wrist Support During Typing: A Controlled, Electromyographic Study." *Applied Ergonomics* 17 (1986): 162–68.

Bergqvist, U. "Video Display Terminals and Health: A Technical and Medical Appraisal of the State of the Art." *Scandinavian Journal of Work Environment and Health* 10, supp. 2 (1984): 1–87.

Bergqvist, U., B. Knave, M. Voss, and R. Wibom. "A Longitudinal Study of VDT Work and Health." *International Journal of Human Computer Interaction* 44 (1992): 197–219.

Bergqvist, U., E. Wolgast, B. Nilsson, and M. Voss. "Musculoskeletal Disorders Among Video Display Terminal Workers: Individual, Ergonomic and Work Organizational Factors." *Ergonomics* 38, no. 4 (1995): 763–76.

Bergqvist, U., E. Wolgast, B. Nilsson, and M. Voss. "The Influence of VDT Work on Musculoskeletal Disorders." *Ergonomics* 38, no. 4 (1995): 754–62.

Berlinguet, L., and D. Berthelette, eds. *Work with Display Units* 1989. Amsterdam: Elsevier Science Publishers, 1990.

Bernstein, A.L., and J.S. Dinesen. "Effect of Pharmacologic Doses of Vitamin B_6 on Carpal Tunnel Syndrome, Electroencephalographic Results, and Pain." *Journal of the American College of Nutrition* 12, no. 1 (1990): 73–76.

Birkbek, M.Q., and T.C. Beer. "Occupations in Relation to the Carpal Tunnel Syndrome." *Rheumatology and Rehabilitation* 4 (1975): 218–21.

Branscum, D. "When it Hurts to Hug." *MacWorld* (October 1989).

Brauninger, U., E. Grandjean, T. Fellmann, and G. Gierer. "Light Characteristics of VDT." *Proceedings of the International Zurich Seminar on Digital Communications*. From seminar held March 9–11 1982.

Browne, C.D., B.M. Nolan, and D.K. Faithfull. "Occupational Repetition Strain Injuries: Guidelines for Diagnosis and Management." *Medical Journal of Australia* 140 (1984): 329–32.

Buchanan, D.A., and D. Boddy. "Advanced Technology and the Quality of Working Life: the Effects of Word Processing on Video Typists." *Journal of Occupational Psychology* 55 (1982): 1–11.

Cain, C., and J. Truck. "VDTs May be Future Work Comp Headache." *Business Insurance* 19 (1985): 18–19.

Cakir, A., D.J. Hart, and T.F.M. Stewart. *Visual Display Terminals*. New York: John Wiley & Sons, Inc., 1980.

Carayon, P., C. Yang, and S. Lim. "Examining the Relationship between Job Design and Worker Strain Over Time in a Sample of Office Workers." *Ergonomics* 38, no. 6 (1995): 1199–1211.

Center for Disease Control. "Working with Video Display Terminals: A Preliminary Health Risk Evaluation." *Morbidity and Mortality Weekly Report* 29 (1980): 307–8.

Chaffin, D.B. "Localized Muscle Fatigue—Definition and Measurement." *Journal of Occupational Medicine* 15, no. 4 (1973): 346–54.

Chaffin, D.B., and G.B.J. Andersson. *Occupational Biomechanics*. New York: John Wiley & Sons, Inc., 1991.

Chattergee, D.S. "Repetition Strain Injury—A Recent Review." *Journal of the Society of Occupational Medicine* 37 (1987): 100–5.

Committee on Cervicobrachial Syndrome in JAIH. "The Report on the Committee in 1972." *Japanese Journal of Industrial Health* 15 (1973): 304–11.

Coray, K., B.F.G. Cohen, and C.S. Piotrkowski. "A Psychophysical Model for Predicting Health Effects of Female Office Workers." In *Work with Computers: Organizational, Management, Stress, and Health Aspects*, edited by M.J. Smith and G. Salvendy, 355–60. Amsterdam: Elsevier Science Publishers, 1989.

Corlett, E.N., J. Wilson, and I. Manenica, eds. *The Ergonomics of Working Posture: Models, Methods and Cases*. London: Taylor & Francis Ltd., 1986.

Cox, T., M. Thirlaway, and S. Cox. "Occupational Well-Being: Sex Differences at Work." *Ergonomics* 27 (1984): 499–510.

Dain, S.J., A.K. McCarthy, and T. Chang-Ling. "Symptoms in VDU Operators." *Am J Optom Physiol Optics* 65 (1988): 162–68.

Dainoff, M.J., A. Happ, and P. Crane. "Visual Fatigue and Occupational Stress in VDT Operators." *Human Factors* 23 (1981): 421–38.

Dainoff, M.J., and M.H. Dainoff. *A Manager's Guide to Ergonomics in the Electronic Office*. New York: John Wiley & Sons, Inc., 1987.

DeKrom, M., A.D.M. Kester, P.G. Knipschild, and F. Spaans. "Risk Factors for Carpal Tunnel Syndrome." *American Journal of Epidemiology* 132 (1990): 1102–10.

Duncan, J., and D. Ferguson. "Keyboard Operating Posture and Symptoms in Operating." *Ergonomics* 17, no. 5 (1974): 651–62.

Elias, R., F. Cail, M. Tisserand, and M. Christmann. "Investigations in Operators Working With CRT Display Terminals; Relationships Between Task Content and Psychophysiological Alterations." In *Ergonomic Aspects of Video Display Terminals*, edited by E. Grandjean and E. Vigliani, 211–18. London: Taylor & Francis Ltd., 1980.

Ellis, D.S. "Speed of Manipulative Performance as a Function of Work Surface Height." *Journal of Applied Psychology* 35 (1951): 289–96.

Fahrbach, P.A., and L.J. Chapman. "VDT Work Duration and Musculoskeletal Discomfort." *American Association of Occupational Health Nurses Journal* 38 (1990): 32–36.

Ferguson, D. "Repetition Injuries in Process Workers." *Medical Journal of Australia* 2 (1971): 408–12.

Ferguson, D., and J. Duncan. "Keyboard Design and Operating Posture." *Ergonomics* 17, no. 6 (1974): 731–44.

Ferguson, D. "The 'New' Industrial Epidemic." *Medical Journal of Australia* 140 (1984): 318–19.

Ferguson, S. "An Australian Study of Telegraphists' Cramp." *British Journal of Industrial Medicine* 28 (1971): 280–85.

Folkers, K., J. Ellis, T. Watanabe, S. Saji, and M. Kaji. "Biochemical Evidence for a Deficiency of Vitamin B6 in the Carpal Tunnel Syndrome Based on a Crossover Clinical Study." *Proceedings of the National Academy of Science, USA* 75 (1978): 3410–12.

Graham, G.J., and C.G. Mills. "Repetitive Strain Injuries." *Medical Journal of Australia* 140 (1984): 380–81.

Grandjean, E. "Ergonomics of VDUs: Review of Present Knowledge." In *Ergonomic Aspects of Video Display Terminals*, edited by E. Grandjean and E. Vigliani, 1–12. London: Taylor & Francis Ltd., 1980.

Grandjean, E. *Fitting the Task to the Man*. 4th ed. London: Taylor & Francis Ltd., 1988.

Grandjean, E., W. Hunting, and M. Pidermann. "VDT Workstation Design: Preferred Settings and Their Effects." *Human Factors* 25 (1983): 161–75.

Grandjean, E. and E. Vigliani, eds. *Ergonomic Aspects of Video Display Terminals*. London: Taylor & Francis Ltd., 1980.

Grandjean, E. "Postural Problems at Office Machine Work Stations." In *Ergonomics and Health in Modern Offices*, edited by E. Grandjean, 445–55. London: Taylor and Francis Ltd., 1984.

Grandjean, E. *Ergonomics in Computerized Offices*. London: Taylor & Francis Ltd., 1987.

Guggenbuhl, U., and H. Krueger. "Musculoskeletal Strain Resulting From Keyboard Use." In *Work with Display Units 1989*, edited by L. Berlinguet and D. Berthelette, 121–28. Amsterdam: Elsevier Science Publishers, 1990.

Hadler, N.M. "Occupational Illness—The Issue of Causality." *Journal of Occupational Medicine* 26 (1984): 587–93.

Hadler, N.M. "Industrial Rheumatology: The Australian and New Zealand Experiences With Arm Pain and Backache in the Workplace." *Medical Journal of Australia* 144 (1986): 191–95.

Hagberg, M., and D.H. Wegman. "Prevalence Rates and Odds Ratios of Shoulder-Neck Diseases in Different Occupational Groups." *British Journal of Industrial Medicine* 44 (1987): 602–10.

Hagberg, M. and G. Sundelin, "Discomfort and Load On the Upper Trapezius Muscle When Operating a Wordprocessor." *Ergonomics* 29 (1986): 1637–45.

Hagberg, M., H. Morgenstern, and M. Keish. "Impact of Occupational and Job Tasks On the Prevalence of Carpal Tunnel Syndrome." *Scandanvian Journal of Work, Environment, and Health* 18 (1922): 337–45.

Hambree, D., and S. Henry. "A Newsroom Hazard Called RSI." *Columbia Journalism Review*, January 1987: 19–24.

Heliovaara, M. "Body Weight, Obesity, and Risk of Herniated Lumbar Invertebral Disc." *Spine* 12 (1987): 469–72.

Hobday, S. "A Keyboard to Increase Productivity and Remove Postural Stress." In *Trends in Ergonomics/Human Factors V*, 1988, edited by F. Aghazadeh, 321–30. Amsterdam: Elsevier Science Publishers, 1988.

Hunting, W., T. Laubli, and E. Grandjean. "Postural and Visual Loads at VDT Workplaces, I. Constrained Postures." *Ergonomics* 24 (1981): 1917–31.

Hunting, W., E. Grandjean, and K. Maeda. "Constrained Postures in Accounting Machine Operators." *Applied Ergonomics* 11 (1980): 145–9.

Hunting, W., T. Laubli, and E. Grandjean. "Constrained Postures of VDU Operators." In *Ergonomic Aspects of Video Display Terminals*, edited by E. Grandjean and E. Vigliani, 175–84. London: Taylor & Francis Ltd., 1980.

Hunting, W., T. Laubli, and E. Grandjean. "Postural and Visual Loads in VDT Workplaces I. Constrained Postures." *Ergonomics* 24, no.12 (1981): 917–31.

Imrhan, S.N. "The Influence of Wrist Position on Different Types of Pinch Strength." *Applied Ergonomics* 22, no. 6 (1991): 379–84.

Imrhan, S.N. "Safety in the Office from Musculoskeletal Stress—VDT Workplace Design." Proceedings of the 10th International System Safety Conference, System Safety Society, Sterling, VA (1991): 6.2–3–1 to 6.2–3–6.

Jeyaratnam, J., C.N. Ong, W.C. Kee, J. Lee, and D. Koh. "Musculoskeletal Symptoms Among VDU Operators." In *Work with Computers: Organizational, Management, Stress, and Health Aspects*, edited by M.J. Smith and G. Salvendy, 330–37. Amsterdam: Elsevier Science Publishers, 1989.

Johansson, G., and G. Aronsson. "Stress Reactions in Computerized Administrative Work." *Journal of Occupational Behavior* 5 (1984): 159–81.

Karasek, R. "Job Demands, Job Decision Latitude, and Mental Strain: Implications for Job Redesign." *Administrative Science Quarterly* 24 (1979): 285–307.

Keegan, J.J. "Alteration of the Lumbar Curve Related to Posture and Seating." *Journal of Bone and Joint Surgery* 35 (1953): 589–603.

Knave, B., R. Wibom, M. Voss, L. Hedstrom, and U. Bergqvist. "Work with Video Display Terminals among Office Employees I: Subjective Symptoms and Discomfort." *Scandinavian Journal of Work, Environment and Health* 11 (1985): 457–66.

Knave, B. and P.G. Wideback, eds. *Work With Display Units* 86. Amsterdam: Elsevier Science Publishers, 1987.

Komoiki, Y., and S. Horiguchi. "Fatigue Assessment on Key Punch Operators, Typists and Others." *Ergonomics* 14 (1971): 101–9.

Konz, S. *Work Design: Industrial Ergonomics*. Worthington, Ohio: Publishing Horizons, Inc., 1990.

Kroemer, K.H.E. "Seating in Plant and Office." *American Industrial Hygiene Association Journal* 32 (1971): 633–52.

Kroemer, K.H.E. "Human Engineering the Keyboard." Human Factors 14 (1972): 51–63.

Kroemer, K.H.E., "Cumulative Trauma Disorders: Their Recognition and Ergonomics Measures to Avoid Them." *Applied Ergonomics* 20, no. 4 (1989): 274–280.

Kumar, S. "Intersubject Varaiation in Biomechanical and Psychophysical Variables with the Use of a Bifocal Computer Desk." In *Advances in Industrial Ergonomics and Safety VII*, edited by A.C. Bittner and P.C. Champney, 1033–39. Philadelphia: Taylor & Francis Ltd., 1995.

Kumar, S. "A Computer Desk for Bifocal Lens Wearers, with Special Emphasis on Selected Telecommunications Tasks." *Ergonomics* 37, no. 10 (1994): 1669–78.

Kumar, S. and W.G.S. Scaife. "A Precision Task, Posture, and Strain." *Journal of Safety Research* 11 (1979): 29–36.

Kuorinka, I., and P. Kosinen. "Occupational Rheumatic Diseases and Upper Limb Strain in Manual Jobs in a Light Mechanical Industry." *Scandinavian Journal of Work, Environment and Health* 6, no. 3 (1979): 39–47.

Kurppa, K., P. Waris, and P. Rokkanen. "Pertendinitis and Tenosynovitis: A Review." *Scandinavian Journal of Work, Environment and Health* 5, no. 3 (1980): 19–24.

Kvarnstrom, S. "Occurence of Musculoskeletal Disorders in a Manufacturing Industry with Special Attention to Occupational Shoulder Disorders." *Scandinavian Journal of Rehabilitation Medicine*, supp. 8 (1983): 1–60.

Lamphier, T.A., C. Crooker, and J. Crooker. "De Quervain's Disease." *Industrial Medicine and Surgery* 34 (1965): 847–56.

Lander, C., G.A. Korbon, D. De Good, and J.C. Rowlingston. "The Balans Chair and Its Semi-Kneeling Position: An Ergonomic Comparison with the Conventional Sitting Position." *Spine*, 12, (1987): 269–72.

Laubli, T., "Preferred Settings in VDT Work: The Zurich Experience." In *Work with Display Units* 86, edited by B. Knave and P.G. Wideback, 249–62. Amsterdam: Elsevier Science Publishers, 1987.

Laubli, T., W. Hunting, and E. Grandjean. "Visual Impairments in VDT Operators Related to Environmental Conditions." In *Ergonomic Aspects of Video Display Terminals*, edited by E. Grandjean and E. Vigliani, 85–94. London: Taylor & Francis Ltd., 1980.

Laubli, T., W. Hunting, and E. Grandjean. "Postural and Visual Loads at VDT Workplaces. II. Lighting Conditions and Visual Impairments," *Ergonomics* (1982): 933–944.

Laville, A., "Postural Reactions Related to Activities on VDU." In *Ergonomic Aspects of Video Display Terminals*, edited by E. Grandjean and E. Vigliani, 167–74. London: Taylor & Francis Ltd., 1980.

Lee, K., N. Swanson, S. Sauter, R. Wickstrom, A. Waikar, and M. Mangum. "A Review of Physical Exercises Recommended for VDT Operators." *Applied Ergonomics* 23, no. 6 (1992): 387–408.

Less, M., W.W.B. Eickelberg, and S. Palgi. "Effects of Work Surface Angles on Productivity Efficiency of Females on a Simple Manual Task." *Perceptual and Motor Skills* 36 (1973): 431–36.

Life, M.A., and S.T. Pheasant. "An Integrated Approach to the Study of Posture in Keyboard Operation." *Applied Ergonomics* 15 (1984): 83–90.

Linton, S., and K. Kamvendo. "Psychosocial Risk Factors for Neck and Shoulder Pain in Secretaries." *Journal of Occupational Medicine* 31 (1989): 609–13.

Lowy, A.H., "Pathogenesis of Overuse Syndromes Affecting the Upper Limb." *Medical Journal of Australia* 2 (1983): 605.

Lundervold, A., "EMG Investigations of Position and Manner of Working in Typewriting." *Acta Phys. Scand.* 24, no. 84 (1956): 124–46.

Luopajarvi, T., I. Kuorinka, M. Viroleinen, and M. Holmberg. "Prevalence of Tenosynovitis and Other Injuries of the Upper Extremities in Repetitive Work." *Scandinavian Journal of Work, Environment and Health* supp. 3 (1979): 48–55.

Maeda, K., "Occupational Cerviobrachial Disorder and Its Causative Factors." *Journal of Human Ergology* 6 (1977): 193–202.

Maeda, K., W. Hunting, and E. Grandjean. "Localized Fatigue in Accounting Machine Operators," *Journal of Occupational Medicine* 22 (1980): 810–16.

Magora, A., "Investigation of the Relation Between Low Back Pain and Occupation V: Psychological Aspects." *Scandinavial Journal of Rehabilitative Medicine* (1973): 191–96.

Mandal, A.C., *The Seated Man: Homo Sedens.* Dafnia, Denmark: 1985.

Martin, D.K., and S.J. Dain. "Postural Modifications of VDU Operators Wearing Bifocal Spectacles." *Applied Ergonomics* 19 (1988): 293–300.

McKay, C., "Work wth Video Display Terminals: Psychosocial Aspects and Health." *Journal of Occupational Medicine* 31, no. 2 (1989): 957–66.

Miller, I., and T.W. Suther. "Display Station Anthropometrics: Preferred Height and Angle Settings of CRT and Keyboard." *Human Factors* 25 (1983): 401–8.

Nachemson, A., and G. Elfstrom. "Intravital Dynamic Pressure Measurements in the Lumbar Discs." *Scandinavian Journal of Rehabilitative Medicine*, supp. 1 (1970): 1–4.

Nakaseko, M., R. Tokunaga, and M. Hosokawa. "History of Occupational Cervicobrachial Disorder in Japan." *Journal of Human Ergology* 11 (1982): 7–16.

Nakaseko, M., E. Grandjean, W. Hunting, and R. Gierer. "Studies on Ergonomic Alphanumeric Keyboards." *Human Factors* 27 (1985): 175–87.

NIOSH—HEW PHS CDC, "The Industrial Environment—Its Evaluation and Control," 1973.

Nishiyama, K., M. Nakaseko, and T. Uehata. "Health Aspects of VDT Operators in the Newspaper Industry." In *Ergonomics and Health in Modern Offices*, edited by E. Grandjean, 113–18. London: Taylor & Francis Ltd., 1984.

O'Brien, B. "Team Technique in RSI Treatment." *Safety in Australia* 8 (1985): 13–14.

Ohara, H., H. Aoyama, and T. Hani. "Health Hazards among Cash Register Operators and the Effects of Improved Working Conditions." *Journal of Human Ergology* 5 (1976): 31–40.

Ohara, H., S. Nakiri, T. Itani, K. Wake, and H. Aoyama. "Occupational Health Hazards Resulting from Elevated Work Rate Situations." *Journal of Human Ergology* 5 (1976): 173–182.

Ong, C.N. "VDT Workplace Design and Physical Fatigue—A Case Study in Singapore." In *Ergonomics and Health in Modern Offices*, edited by E. Grandjean, 484–89. London: Taylor & Francis Ltd., 1984.

Onishi, N., H. Nomura, and K. Sakai. "Fatigue and Strength of Upper Limb Muscles of Flight Reservation System Operators." *Journal of Human Ergology* 2 (1973): 133–41.

Ostberg, O. "Accommodation and Visual Fatigue in Display Work." In *Ergonomic Aspects of Video Display Terminals*, edited by E. Grandjean and E. Vigliani, 41–52. London: Taylor & Francis Ltd., 1980.

Oxford, H.W. "Anthropometric Data for Educational Chairs." *Ergonomics* 12 (1969): 140–61.

Putz-Anderson, V., ed. *Cumulative Trauma Disorders: A Manual for Musculoskeletal Diseases of the Upper Limbs*. London: Taylor & Francis Ltd., 1988.

Rose, L. "Workplace Video Display Terminals and Visual Fatigue." *Journal of Occupational Medicine* 29, no. 4 (1985): 321–24.

Rose, M. "Death of the QWERTY Keyboard?" *Design World* 8 (1985): 36–43.

Rose, M.J. "Keyboard Operating Posture and Actuation Force: Implications for Muscle Over-use." *Applied Ergonomics* 22 (1991): 198–203.

Sanders, M.S., and E.J. McCormick. *Human Factors in Engineering and Design*. 6th ed. New York: McGraw-Hill Book Co., 1987.

Sauter, S.L., L.M. Schleiffer, and S.J. Knutson. "Work Posture, Workstation Design, and Musculoskeletal Discomfort in a VDT Data Entry Task." *Human Factors* 33 (1991): 151–67.

Sauter, S.L., Gottlieb, M.S., K.C. Jones, V.N. Dodson, and K.M. Rohrer. "Job and Health Implications of VDT Use: Initial Results of the Wisconsin-NIOSH Study." *Communications of the ACM* 26 (1983): 284–94.

Sauter, S.L. and N.G. Swanson. "The Effects of Frequent Rest Breaks on Performance and Well-being in Repetitive VDT Work." In *Work with Display Units '92*, edited by H. Luczak, A.E. Cakir and G. Cakir, D52. Berlin: Technische Universitat Berlin, 1992.

Sauter, S.L., L.J. Chapman, and S.J. Knutson. *Improving VDT Work: Causes and Control of Health Concerns in VDT Use*. Lawrence, Kansas: Report Store, 1985.

Scarf, G.E., and D. Wilcox. "Alleged Work Related Injuries." *Medical Journal of Australia* 141 (1984): 765.

Sheedy, J.E. "Video Display Terminals—Solving the Vision Problems." In *Problems in Optometry: Environmental Optics*, edited by J.E. Sheedy, 1–16. Philadelphia: J.B. Lippincott, 1990.

Sheilds, R.K., and T.M. Cook. "Effect of Seat Angle and Lumbar Support on Seated Buttock Pressure." *Journal of the American Physical Therapy Association* (1988): 1682–86.

Silverstein, B.A., L.J. Fine, and T.J. Armstrong. "Occupational Factors and Carpal Tunnel Syndrome." *American Journal of Industrial Medicine* 11 (1987): 343–58.

Smith, M.J., B.G. Cohen, L.W. Stammerjohn, and A. Happ. "An Investigation of Health Complaints and Job Stress in Video Disply Operations." *Human Factors* 23, no. 4 (1981): 387–400.

Smith, M.J. "Job Stress in Video Display Operations." In *Ergonomic Aspects of Video Display Terminals*, edited by E. Grandjean and E. Vigliani, 201–10. London: Taylor & Francis Ltd., 1980.

Smith, M.J., P. Carayon, K.J. Sanders, S.Y. Lin, and D. LeGrande. "Employee Stress and Complaints in Jobs with and without Electronic Performance Monitoring." *Applied Ergonomics* 23 (1992): 17–27.

Smith, M.J. and G. Salvendy., eds. *Work with Computers: Organizational, Management, Stress, and Health Aspects*, Amsterdam: Elsevier Science Publishers, 1989.

Stammerjohn, L.W., M.J. Smith, and B.G.F. Cohen. "Evaluation of Work Station Design Factors in VDT Operations." *Human Factors* 23, no. 4 (1981): 401–12.

Starr, S., C. Thompson, and S. Shute. "Effects of Video Display Terminals on Telephone Operators." *Human Factors* 24 (1982): 699–711.

Starr, S.J., S.J. Shute, and C.R. Thompson. "Relating Posture to Discomfort in VDT Use." *Journal of Occupational Medicine* 27 (1985): 269–71.

Stock, S.R. "Workplace Ergonomic Factors and the Development of Musculoskeletal

Disorders of the Neck and Upper Limbs: A Meta-analysis." *American Journal of Industrial Medicine* 19 (1991): 87–107.

Stone, W.E. "Repetitive Strain Injuries." *Medical Journal of Australia* 2 (1983): 616–18.

Sundelin, G., and M. Hagberg. "The Effects of Different Pause Types on Neck and Shoulder EMG Activity During VDU Work." *Ergonomics* 32 (1989): 527–37.

Tichauer, E.R. "Ergonomic Aspects of Biomechanics." In NIOSH—HEW PHS CDC (1973), 431–92.

U.S. Department of Labor. Bureau of Labor Statistics. *BLS Reports on Survey of Occupational Injuries and Illnesses in 1977–1990*. Washington, D.C., 1991.

VanWely, P. "Design and Disease." *Applied Ergonomics* 1 (1970): 262–69.

Vassidieff, A., and S. Dain. "Bifocal Wearing and VDU Operation: A Review and Graphic Analysis." *Applied Ergonomics* 17 (1986): 82–86.

VDT News. "Lead Aprons for Operators: Help or Hindrance." *VDT News* 1, no. 4 (1984).

Waersted, M., R.A. Bjorklund, and R.H. Westgaard. "Shoulder Muscle Tension Induced by Two VDU-based Tasks of Different Complexity." *Ergonomics* 34 (1991): 137–50.

Wall, T.D., B. Burnes, C.W. Clegg, and N.J. Kemp. "New Technology, Old Jobs." *Work and People* 10 (1984): 15–21.

Weber, A., E. Sancin, and E. Grandjean. " Effects of Various Keyboard Heights on EMG and Physical Discomforts." In *Ergonomics and Health in Modern Offices*, edited by E. Grandjean, 477–83. London: Taylor & Francis Ltd., 1984.

Welch, R. "The Causes of Tenosynovitis in Industry." *Industrial Medicine* 41 (1972): 16–19.

Wells, M.J. "Industrial Incidence of Soft Tissue Syndromes." *Physical Therapy Review* 41 (1961): 512–15.

Westgaard, R.H., A. Aaras, M. Waersted, and T. Jensen. "Muscle Load and Illness Associated With Constrained Body Postures." In *The Ergonomics of Working Posture: Models, Methods and Cases*, edited by E.N. Corlett, J. Wilson, and I. Manenica, 5–18. London: Taylor & Francis Ltd., 1986.

Westgaard, R.H., and R. Bjorklund. "Generation of Muscle Tension Additional to Postural Muscle Load." *Ergonomics* 30 (1987): 911–23.

Zipp, P., E. Haider, N. Halpern, and W. Rohmert. "Keyboard Design through Physiological Strain Measurements." *Applied Ergonomics* 14 (1983): 117–22.

INDEX